# AS/A-LEVEL YEAR 1

## STUDENT GUIDE

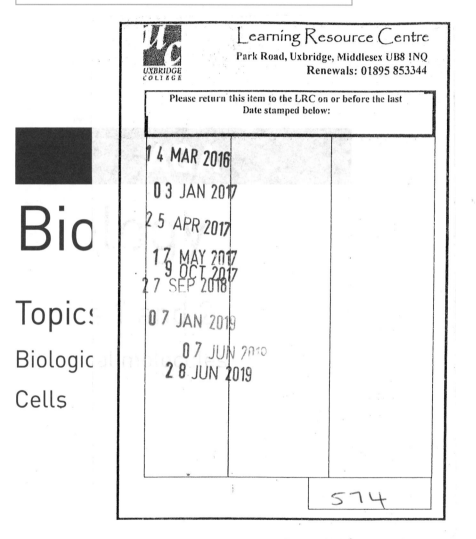
# Bio

## Topics

Biologic

Cells

Pauline Lowrie

PHILIP ALLAN FOR
HODDER
EDUCATION
AN HACHETTE UK COMPANY

Philip Allan, an imprint of Hodder Education, an Hachette UK company, Blenheim Court, George Street, Banbury, Oxfordshire OX16 5BH

*Orders*

Bookpoint Ltd, 130 Milton Park, Abingdon, Oxfordshire OX14 4SB

tel: 01235 827827

fax: 01235 400401

e-mail: education@bookpoint.co.uk

Lines are open 9.00 a.m.–5.00 p.m., Monday to Saturday, with a 24-hour message answering service. You can also order through the Hodder Education website: www.hoddereducation.co.uk

© Pauline Lowrie 2015

ISBN 978-1-4718-4324-2

First printed 2015

Impression number 5 4 3 2 1

Year 2018 2017 2016 2015

This guide has been written specifically to support students preparing for the AQA AS and A level Biology (Topics 1 and 2) examinations. The content has been neither approved nor endorsed by AQA and remains the sole responsibility of the author.

Cover photo: Sergey Nivens/Fotolia

Typeset by Integra Software Services Pvt Ltd, Pondicherry, India

Printed in Italy

Hachette UK's policy is to use papers that are natural, renewable and recyclable products and made from wood grown in sustainable forests. The logging and manufacturing processes are expected to conform to the environmental regulations of the country of origin.

# Contents

# Content Guidance

# Questions & Answers

# ■ Getting the most from this book

## Exam tips

Advice on key points in the text to help you learn and recall content, avoid pitfalls, and polish your exam technique in order to boost your grade.

## Knowledge check

Rapid-fire questions throughout the Content Guidance section to check your understanding.

## Knowledge check answers

**1** Turn to the back of the book for the Knowledge check answers.

## Summaries

■ Each core topic is rounded off by a bullet-list summary for quick-check reference of what you need to know.

---

**Exam-style questions**

**Commentary on the questions**

Tips on what you need to do to gain full marks, indicated by the icon ⓔ

**Sample student answers**

Practise the questions, then look at the student answers that follow.

**Commentary on sample student answers**

Find out how many marks each answer would be awarded in the exam and then read the comments (preceded by the icon ⓔ) showing exactly how and where marks are gained or lost.

---

### Questions & Answers

**(i)** Complete Table 2 to show the figures for *E. coli*. Explain how you arrived at this answer. [1 mark]

**(ii)** The structure of the DNA in the virus is not the same as the structure of DNA in the other organisms. Suggest what this difference in DNA structure might be. [1 mark]

ⓔ This question is testing your knowledge and understanding. It starts with simple recall but then moves to giving you information that is probably new to you, to check whether you can apply your knowledge to an unfamiliar situation.

**Student A**

**(a) (i)** (cross placed right next to the line after telophase) ✓

**(ii)** DNA replication and protein synthesis ✓

ⓔ 2/2 marks awarded Student A clearly understands this topic well.

**(b)** Mitosis will not be able to go past prophase ✓, because spindle fibres are needed for the chromosomes to line up on during metaphase ✓.

ⓔ 2/2 marks awarded Student A says clearly what will happen and then gives a correct reason to explain.

**(c) (i)**

| Source of DNA | Adenine % | Guanine % | Thymine % | Cytosine % |
|---|---|---|---|---|
| Human | 30 | 20 | 30 | 20 |
| Turtle | 28 | 22 | 28 | 22 |
| *E. coli* | 24 | 26 | 24 | 26 |
| Salmon | 29 | 21 | 29 | 21 |
| Virus | 25 | 24 | 33 | 18 |

This is because adenine pairs with thymine, and cytosine with guanine so the amount of adenine nucleotides will be the same as thymine. The rest of the nucleotides will be half cytosine and half guanine ✓.

**(ii)** The virus DNA must be single-stranded because A does not equal T and C does not equal G ✓

ⓔ 2/2 marks awarded This is a precise answer, for full marks.

**Student B**

**(a) (i)** (a cross in telophase) ✗ **(ii)** DNA replication and respiration ✓

ⓔ 1/2 marks awarded No mark is awarded for (a)(i) because cytokinesis happens immediately *after* telophase. However, student B gets a mark for two correct processes in (a)(ii).

**(b)** Metaphase will not happen ✓ because a spindle is needed for the chromosomes to attach to ✓.

ⓔ 2/2 marks awarded This is all that is needed for 2 marks.

76 AQA Biology

# ■About this book

This guide will help you to prepare for AQA AS/A-Level Biology Year 1, topics 1 and 2. At AS, topics 1 and 2 form half the content of papers 1 and 2. At A-level these topics are examined in paper 1 (together with topics 3 and 4) and in paper 3 (together with topics 3–8).

The **Content Guidance** section covers all the facts you need to know and concepts you need to understand for topics 1 and 2. It is really important that you focus on *understanding* and not just learning facts, as the examiners will be testing your ability to apply what you have learned in new contexts. This is impossible to do unless you really understand everything. The Content Guidance also includes exam tips and knowledge checks to help you prepare for your exams.

The **Questions & Answers** section shows you the type of questions you can expect in the exam. It would be impossible to give examples of every kind of question in one book, but these should give you a flavour of what to expect. Two students, student A and student B, have attempted each question. Their answers, and the accompanying comments, should help you to see what you need to do to score a good mark — and how you can easily *not* score a mark even though you probably understand the biology.

## What can I assume about the guide?

You can assume that:

- the basic facts you need to know and understand are stated explicitly
- the major concepts you need to understand are explained clearly
- the questions at the end of the guide are similar in style to those that will appear in the final examination
- the questions assess the different assessment objectives
- the standard of the marking is broadly equivalent to that which will be applied to your answers

## How should I use this guide?

The guide lends itself to a number of uses throughout your course — it is not *just* a revision aid. You could:

- use it to check that your notes cover the material required by the specification
- use it to identify your strengths and weaknesses
- use it as a reference for homework and internal tests
- use it during your revision to prepare 'bite-sized' chunks of related material, rather than being faced with a file full of notes

You could use the Questions & Answers section to:

- identify the terms used by examiners and show what they expect of you
- familiarise yourself with the style of questions you can expect
- identify the ways in which students gain, or fail to gain, marks

# Develop your examination strategy

Just as reading the *Highway Code* alone will not help you to pass your driving test, this guide cannot help to make you a good examination candidate unless you develop and maintain all the skills that examiners will test in the final exams. You also need to be aware of the type of questions examiners ask and where to find them in the exams. You can then develop your own personal examination strategy. But, be warned, this is a highly personal and long-term process; you cannot do it a few days before the exam.

## Things you *must* do

- Clearly, you must know some biology. If you don't, you cannot expect to get a good grade. This guide provides a succinct summary of the biology you must know.
- Be aware of the skills that examiners *must* test in the exams. These are called assessment objectives and are described in the AQA Biology specification.
- Understand the weighting of the assessment objectives that will be used. These are as follows.

| Assessment objective | Brief summary | Marks in A-level paper 1/% | Marks in A-level paper 2/% | Marks in A-level paper 3/% | Marks in AS paper 1/% | Marks in AS paper 2/% |
|---|---|---|---|---|---|---|
| AO1 | Knowledge and understanding | 44–48 | 23–27 | 28–32 | 47–51 | 33–37 |
| AO2 | Application of knowledge and understanding | 30–34 | 52–56 | 35–39 | 35–39 | 41–45 |
| AO3 | Analyse, interpret and evaluate scientific information, ideas and evidence | 20–24 | 19–23 | 31–35 | 13–17 | 21–25 |

- Use past questions and other exercises to develop all the skills that examiners must test. Once you have developed them all, keep practising to maintain them.
- Understand where in your exams different types of questions occur. For example, the final question on AS paper 2 will always be worth 10 marks and will test AO1 by requiring you to write extended prose. If that is the skill in which you feel most comfortable, and many AS students do, why not attempt this question first?
- Remember that mathematical skills account for about 10% of the marks. Do make sure you can carry out these calculations, including percentages, ratios and rates of reaction.
- You need to be familiar with the techniques you have learned in the required practicals, and be able to describe how these techniques might be used in a different context. Also, you need to be able to evaluate practical investigations and data presented to you in the exam. Answers to the questions set in the required practicals are given on page 91.

# Content Guidance

# ■ Biological molecules

## Monomers and polymers

**Monomers** are small molecules that form the building blocks of larger molecules. **Polymers** are large molecules formed when many similar smaller molecules, or monomers, join together.

Two monomers join together by a **condensation** reaction. This means that when the monomers join together a molecule of water is removed. You can see this in Figure 1.

A polymer can be broken down again to monomers by a **hydrolysis** reaction. This is when a molecule of water is added to break the bond joining the molecules together.

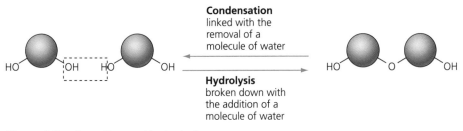

**Figure 1** Condensation and hydrolysis

## Carbohydrates

### Monosaccharides

The simplest carbohydrates are **monosaccharides**. The monosaccharide that you need to know most about is glucose. It has the formula $C_6H_{12}O_6$. There are two kinds of glucose, as shown in Figure 2. These two different forms of glucose are called isomers.

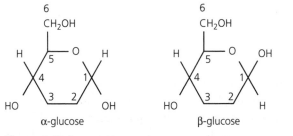

**Figure 2** Alpha- and beta-glucose

Galactose and fructose are also monosaccharides. They too have the formula $C_6H_{12}O_6$.

**Exam tip**

You need to be able to draw glucose but you do not need to memorise the exact difference between alpha- and beta-glucose. All you need to remember is that the H and OH are arranged differently at the right-hand side of the molecule.

**Knowledge check 1**

Galactose and fructose have the same formula as glucose $(C_6H_{12}O_6)$. Explain how they can be different monosaccharides from glucose.

## Disaccharides

**Disaccharides** are formed when two monosaccharides join together by a condensation reaction (Figure 3). The bond formed is a glycosidic bond.

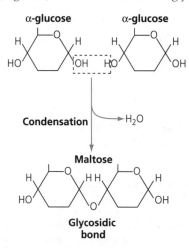

**Figure 3** Maltose is formed when two molecules of glucose join together

Table 1 shows the monosaccharides that make up some different disaccharides.

| Name of disaccharide | Monsaccharides it contains |
| --- | --- |
| Maltose | Glucose + glucose |
| Sucrose | Glucose + fructose |
| Lactose | Glucose + galactose |

**Table 1** Composition of some different disaccharides

## Polysaccharides

**Polysaccharides** are polymers made of many different monosaccharides joined together. You need to know starch, glycogen and cellulose particularly well. These are all polymers of glucose. Glycogen and starch are both polymers of alpha-glucose. They are used as storage carbohydrates, but glycogen is found in animal cells while starch is found in plant cells. Structurally they are similar, although glycogen is more branched. They both have similar properties:

- They are insoluble, so they do not affect osmosis.
- They are branched, so there are lots of 'ends' from which to break off glucose molecules when needed.
- They coil up, making them compact, so a lot of glucose can be stored in a small space.

Cellulose is a polymer of beta-glucose. This gives it a twisted glycosidic bond, which cannot be broken down by enzymes in mammalian guts. It is used to form plant cell walls. It has some important properties that make it ideal as a structural polysaccharide:

- It forms long, straight chains that are difficult to stretch.
- Hydrogen bonds form between molecules, forming fibrils with a great deal of tensile strength.
- It is insoluble, so it cannot 'wash away'.

# Lipids

Lipids are made of the elements carbon, hydrogen and oxygen, the same as carbohydrates, but they contain more hydrogen and less oxygen than carbohydrates. The main lipids that you need to know about are triglycerides, which is the chemical name for fats and oils.

## Triglycerides

Triglycerides are made up of glycerol and fatty acids. Their structure is shown in Figures 4 and 5.

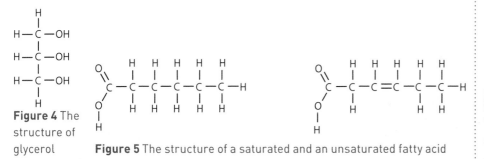

**Figure 4** The structure of glycerol

**Figure 5** The structure of a saturated and an unsaturated fatty acid

**Exam tip**

The shorthand way to show the structure of a fatty acid is R–COOH, where R represents the fatty acid 'tail'.

Notice that an unsaturated fatty acid has at least one double carbon=carbon bond in the hydrocarbon 'tail'. A saturated fatty acid has all single C–C bonds.

A triglyceride is formed when three fatty acids join to a glycerol molecule by a condensation reaction (Figure 6). The bonds formed are called ester bonds.

Note that a triglyceride is not a polymer because it is made up of only four components, rather than a large number of similar, smaller molecules.

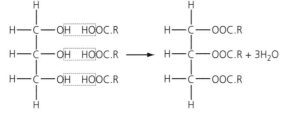

Glycerol          Fatty acids

**Figure 6** Formation of a triglyceride

## Phospholipids

**Phospholipids** are similar to triglycerides, except that one of the fatty acids is replaced by a phosphate group. The structure of a phospholipid is shown in Figure 7.

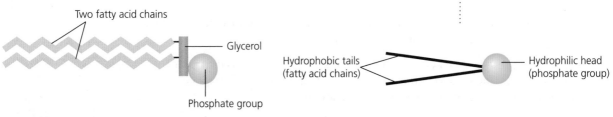

**Figure 7** The structure of a phospholipid

The key feature of a phospholipid is that the glycerol–phosphate 'head' of the molecule is soluble in water, so it is hydrophilic or 'water-loving'. The fatty acid 'tail' is insoluble in water, so it is hydrophobic or 'water-hating'. This means that the molecule is polar, as one end of the molecule has different properties from the other end. It is important in the structure of cell membranes.

# Proteins

## General properties of proteins

### Amino acids

Proteins are polymers of amino acids. There are twenty different amino acids, but they all have the same basic structure shown in Figure 8.

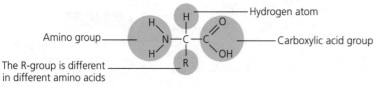

**Figure 8** The structure of an amino acid

**Exam tip**

You can remember the elements that make up an amino acid by the nonsense word 'CHONS'.

Amino acids join together by a condensation reaction to form dipeptides (Figure 9).

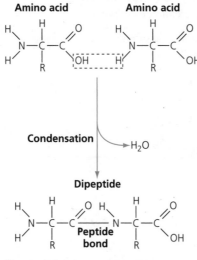

**Figure 9** Joining amino acids

### Polypeptides and proteins

More amino acids join on to form polypeptides. The **primary structure** of a protein is the order of amino acids in the polypeptide chain.

Hydrogen bonds form between different amino acids in the chain. This causes the chain to fold or coil and produce different shapes, such as an alpha-helix or a beta-pleated sheet. This is the protein's **secondary structure**.

The whole polypeptide then coils and folds into an overall three-dimensional shape. This is held in place by weak hydrogen or ionic bonds, or stronger disulfide bridges between the R-groups of sulfur-containing amino acids. This is the protein's **tertiary structure**.

Some proteins are made of more than one polypeptide. The overall shape this takes is called the **quaternary structure**.

When a protein is exposed to high temperatures or a large change in pH, the ionic and hydrogen bonds holding the protein in its tertiary structure break. This changes the shape of the protein, and the protein is said to be **denatured**. The shape of a protein is usually very important to its function, so once a protein is denatured, it tends to lose its function.

**Knowledge check 2**

Name the bonds that hold a protein's secondary structure together.

**Knowledge check 3**

Name the bonds formed **a** between glycerol and a fatty acid and **b** between two amino acids.

## Tests for molecules in food

The tests for molecules in food are shown in Table 2.

| Substance | Test | Brief details of test | Positive result |
|---|---|---|---|
| Protein | Biuret test | Add sodium hydroxide to the test sample Add a few drops of dilute copper sulfate solution | Solution turns mauve |
| Carbohydrates: Reducing sugars | Benedict's test | Heat test sample with Benedict's reagent | Orange-red precipitate is formed |
| Carbohydrates: Non-reducing sugars | Benedict's test | Check that there is no reducing sugar present by heating part of sample with Benedict's reagent (see above) Hydrolyse rest of sample by heating with dilute hydrochloric acid Neutralise by adding sodium hydrogen carbonate Test sample with Benedict's solution | Orange-red precipitate is formed |
| Starch | Iodine test | Iodine/potassium iodide solution | Turns blue-black |
| Lipid | Emulsion test | Dissolve the test sample by shaking with ethanol Pour the resulting solution into water in a test tube | A white emulsion is formed |

**Table 2** Tests for carbohydrates, proteins and lipids

## Summary

- Condensation reactions join two molecules together by removing a molecule of water. Hydrolysis reactions break bonds by adding a molecule of water.
- Proteins are polymers of amino acids, which join together to form peptide bonds.
- Triglycerides are formed when three fatty acids bind to a glycerol molecule.
- Polysaccharides are polymers of monosaccharides.

## Many proteins are enzymes

### How enzymes work

Enzymes work by lowering the **activation energy** needed for a reaction to take place. Activation energy is the energy needed to start a reaction (Figure 10).

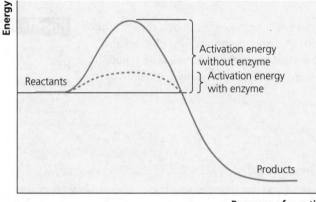

**Figure 10** Activation energy

This means that reactions in the body can take place at low temperatures and atmospheric pressure. Enzymes can do this because they bring the reacting molecules close together. This is explained by the **induced fit** model (Figure 11).

Enzymes are proteins so they have a complex tertiary structure. Part of their shape is an **active site** into which the substrate fits. As the substrate binds to the active site, the active site changes shape so that it becomes complementary to the substrate. Once the enzyme–substrate complex is formed, the enzyme catalyses the reaction and the products are released. The enzyme can then be used over and over again.

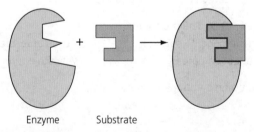

**Figure 11** The induced fit model of enzyme action

### Factors that affect enzyme activity

### Temperature

At low temperatures, enzyme activity is low because the substrate and enzyme have too little kinetic energy, so there are not enough molecular collisions for enzyme–substrate complexes to form. As temperature increases, so does the rate of reaction. However, once a certain point is reached, the enzyme vibrates so vigorously that some of the weak bonds holding it in its tertiary structure break. This means that the enzyme's active site is no longer the right shape for the substrate to fit in. The enzyme has been **denatured**.

**Knowledge check 4**

A student wrote in an exam that an enzyme's active site is the same shape as the substrate. Why was this a mistake?

**Knowledge check 5**

Explain why sucrase will hydrolyse sucrose but will not hydrolyse maltose.

Each enzyme has an **optimum temperature** at which the rate of reaction occurs at its fastest. For human enzymes, the optimum temperature is usually 37°C, but this is not the case for all enzymes. You can see this in Figure 12.

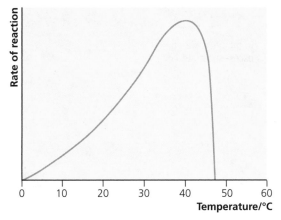

**Figure 12** Optimum temperatures for enzymes

**Knowledge check 6**

Name the bonds that break when an enzyme denatures.

## pH

The pH of a solution is a measure of its hydrogen ion concentration. The higher the concentration of hydrogen ions, the lower the pH, and the more acid the solution. Every enzyme has an **optimum pH** at which the rate of reaction is at its fastest. This is the pH at which the enzyme's active site is the perfect shape to fit with the substrate. As the pH becomes higher or lower than this, two changes occur:

■ The charges on the amino acids in the enzyme's active site change so that the substrate no longer fits.
■ The hydrogen and ionic bonds that hold the enzyme in its tertiary structure are broken, denaturing the enzyme.

This is shown in Figure 13. Remember that, while many enzymes have an optimum pH of 7, this is not the case for all enzymes.

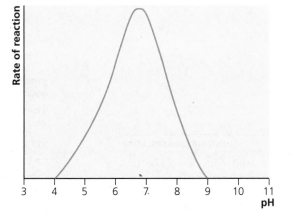

**Figure 13** The effect of pH on enzymes

## Enzyme concentration

If there is an unlimited amount of substrate, then adding more enzyme increases the rate of reaction (Figure 14).

**Knowledge check 7**

A student wanted to find the optimum temperature of amylase. She put the same volume and concentration of amylase and starch into each of several tubes, and changed the temperature using a water bath. She also added a buffer solution to each tube. Why was this necessary?

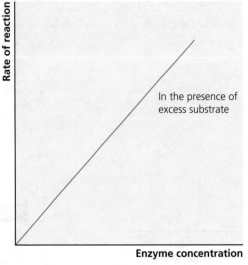

**Figure 14** The effect of adding more enzyme on rate of reaction

This is because adding more enzyme means there are more active sites available, so the rate of reaction increases.

### Substrate concentration

If the concentration of substrate is increased, the rate of reaction changes (Figure 15).

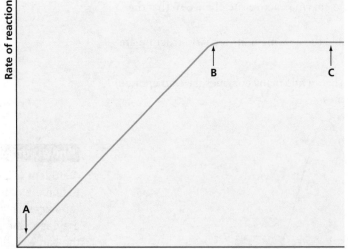

**Figure 15** The effect of substrate concentration on rate of reaction

At A, the rate of reaction is very low. This is because there are plenty of enzymes available, but few of them are being used, as there are not enough substrate molecules. So many of the enzyme active sites are unfilled. As the substrate concentration increases between A and B, the rate of reaction increases (i.e. more product is formed per minute). This is because more of the enzyme's active sites are being filled. However, at B, the rate of reaction stays at a constant high rate despite more substrate being added between B and C. This is because all the enzyme's

> **Knowledge check 8**
>
> What shape would the graph in Figure 14 be if the substrate concentration was not unlimited?

> **Knowledge check 9**
>
> The graph in Figure 15 shows the rate of reaction at the enzyme's optimum temperature of 37°C. Copy the graph and sketch a line on it to show the shape of the graph if the reaction had been carried out at 25°C.

active sites are being used, over and over again, so adding more substrate makes no difference. (It is as if the substrates are having to 'queue up' to get to an active site.)

## Competitive inhibitors

**Inhibitors** are molecules that slow down the rate of an enzyme-controlled reaction. Competitive inhibitors are molecules that are similar in shape to the substrate. This means that the inhibitor can fit into the enzyme's active site. However, because it isn't the right substrate, no reaction can happen. The inhibitor simply blocks the active site and stops the enzyme from binding with any substrates until the inhibitor comes out of the active site again. You can see how competitive inhibitors work in Figure 16.

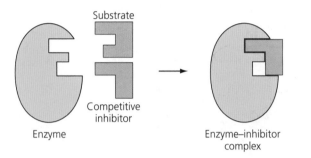

**Figure 16** Competitive inhibition

Figure 17 shows the effect of a competitive inhibitor on the rate of an enzyme-controlled reaction. You will see that, at low substrate concentrations, the inhibitor slows down the rate of reaction a lot. However, as more substrate is added, the inhibitor has less and less of an effect. This is because when there is more substrate compared with inhibitor, the chances of a substrate molecule colliding with the enzyme's active site is greater than the chance of an inhibitor entering the active site. As more and more substrate is added, this effectively dilutes the inhibitor so, once a high concentration of substrate is present, the inhibitor has almost no effect on the rate of reaction.

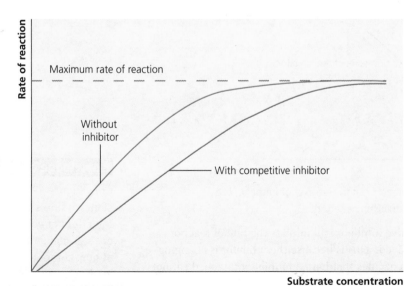

**Figure 17** The effect of a competitive inhibitor on the rate of an enzyme-controlled reaction

**Knowledge check 10**

Copy the graph in Figure 17 and sketch a line to show what would happen if an even greater concentration of competitive inhibitor was added.

**Knowledge check 11**

Alcohol is broken down in the body to acetaldehyde. This would cause a person to feel nauseous if it accumulated in the body, but normally it is broken down quickly by the enzyme aldehyde oxidase. A drug called disulfiram is a competitive inhibitor of aldehyde oxidase. It is sometimes used as a drug to help people overcome their drinking habit. Explain how.

### Non-competitive inhibitors

Non-competitive inhibitors don't fit into the active site of the enzyme, but bind to a different point on the enzyme. However, they change the shape of the enzyme's active site so that the substrate can no longer fit in (Figure 18).

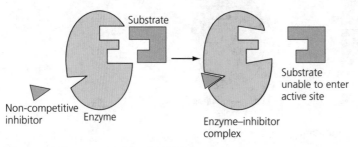

**Figure 18** Non-competitive inhibition

Figure 19 shows the effect of a non-competitive inhibitor on the rate of an enzyme-controlled reaction.

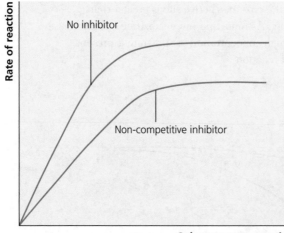

**Figure 19** The effect of a non-competitive inhibitor on the rate of an enzyme-controlled reaction

You will notice that a non-competitive inhibitor still inhibits the rate of reaction no matter how much substrate is added. This is because the inhibitor is changing the shape of the active site, so any enzymes that have an inhibitor attached cannot combine with a substrate molecule.

**Knowledge check 12**

Copy Figure 19 and sketch a line showing the effect of adding a greater concentration of non-competitive inhibitor.

## Required practical 1

### Investigating factors that affect enzyme activity

#### Investigating catalase activity in potato

Potato tissue contains an enzyme called catalase. This enzyme breaks down hydrogen peroxide into water and oxygen as shown in the equation:

$$2H_2O_2 \xrightarrow{\text{catalase}} 2H_2O + O_2$$

A student set up the apparatus shown in Figure 20. The inverted test tube was completely filled with water at the start and the conical flask was empty.

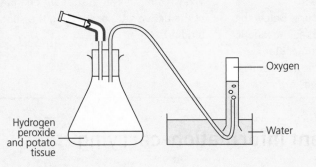

**Figure 20**

The student cut five discs of potato tissue and placed them into the conical flask. She replaced the bung. Next she measured $5\,cm^3$ of hydrogen peroxide solution into the syringe. She pushed the plunger on the syringe so that the hydrogen peroxide entered the flask and started a stop clock at the same time. She measured the volume of oxygen given off every 20 seconds. Table 3 shows her results.

| Time from adding hydrogen peroxide to potato tissue/s | Volume of oxygen gas/cm$^3$ |
|:---:|:---:|
| 0 | 0 |
| 20 | 4.3 |
| 40 | 6.5 |
| 60 | 7.7 |
| 80 | 8.4 |
| 100 | 8.8 |
| 120 | 9.1 |
| 140 | 9.3 |

**Table 3**

1 Give a suitable control for this investigation.
2 How could the student calculate the rate of reaction (a) over the first 20 seconds and (b) in the last 20 seconds?
3 How could the student adapt this investigation to find the optimum temperature of catalase?
4 How could the student adapt this investigation to find the effect of substrate concentration on enzyme activity?
5 The student should have repeated the investigation several times and found the mean values. Explain why.

Turn to page 91 for the answers.

## Summary

- Enzymes lower the activation energy needed for reactions to occur, enabling them to take place at modest temperatures and atmospheric pressure.
- Enzymes are proteins with a specific tertiary structure. They have an active site into which the substrate fits.
- When the substrate enters the active site, the active site changes shape to become complementary to the substrate.
- Enzymes have an optimum temperature. Below this point, increasing temperature increases the kinetic energy of molecules and increases molecular collisions. Above the optimum, the enzyme's active site changes shape and it is denatured.
- Enzymes have an optimum pH. Above or below this point, the enzyme's active site changes shape and the enzyme becomes denatured.
- Enzyme activity is also affected by enzyme concentration and substrate concentration.
- Competitive inhibitors fit into the enzyme's active site, preventing the substrate from binding.
- Non-competitive inhibitors fit into the enzyme away from the active site, but change the shape of the active site, preventing the substrate from binding.

# Nucleic acids: important information-carrying molecules

## Structure of DNA and RNA

### DNA

Deoxyribonucleic acid (DNA) is an important molecule in cells because it stores genetic information. It is a polymer made of many nucleotides joined together. The structure of a nucleotide is shown in Figure 21.

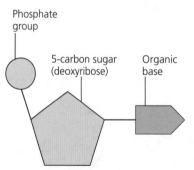

**Figure 21** The structure of a nucleotide

A nucleotide has three components:

- a pentose (or 5-carbon sugar) called **deoxyribose**
- a phosphate group
- a nitrogen-containing organic base, which can be adenine, guanine, cytosine or thymine

Two nucleotides join together by a **condensation reaction**, forming a **phosphodiester** bond. More nucleotides join on to this to form a **polynucleotide** strand. You can see this in Figures 22 and 23.

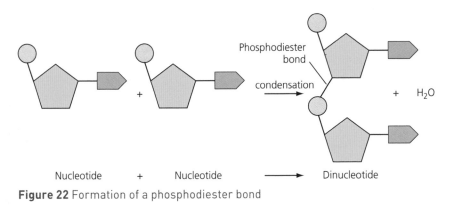

**Figure 22** Formation of a phosphodiester bond

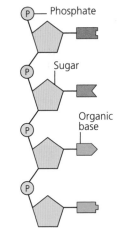

**Figure 23** Formation of a polynucleotide strand

Two polynucleotide strands fit together to make a molecule of DNA, one strand being upside down compared with the other. However, they way the strands fit together is very specific. Two nucleotides with complementary bases (one from each strand) fit together by hydrogen bonds, which join the two strands together. Base pairing only occurs between complementary bases. Adenine always pairs with thymine, and guanine always pairs with cytosine (Figure 24).

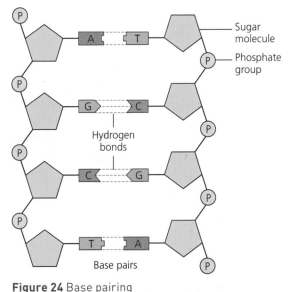

**Figure 24** Base pairing

These two strands then twist up to form a double helix shape (Figure 25).

**Exam tip**

Although hydrogen bonds are relatively weak, there are so many of them in a molecule of DNA that they are collectively fairly strong.

**Knowledge check 13**

20% of the nucleotides in a piece of DNA contain thymine. What percentage of nucleotides contain guanine?

**Exam tip**

You need to remember the base pairing rules — that A pairs with T and C pairs with G. An easy way to remember this is to think of people you know with the initials AT or TA, and CG or GC.

**Figure 25** DNA double helix

### RNA

Ribonucleic acid (RNA) is another kind of nucleic acid. Like DNA it is a polymer made up of nucleotides, but its nucleotides are slightly different. You can see the structure of an RNA nucleotide in Figure 26.

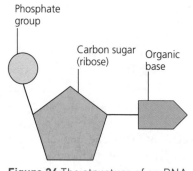

**Figure 26** The structure of an RNA nucleotide

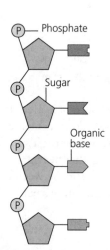

**Figure 27** The structure of RNA

RNA nucleotides, like DNA nucleotides, contain the three nitrogen-containing organic bases adenine, cytosine and guanine, but they never have thymine. Instead, they have nucleotides containing a different organic base called uracil. The nucleotides join together to form a polynucleotide strand, but RNA stays as a single polynucleotide strand rather than forming a double helix. Figure 27 shows the structure of RNA.

RNA molecules are relatively short. RNA transfers genetic information from the DNA in the chromosomes to the ribosomes. Ribosomes are made of RNA and protein.

## DNA replication

DNA replication is described as **semi-conservative** (Figure 28) because one strand of the old molecule remains intact and a new strand is synthesised following base-pairing rules. This happens before a cell can divide. It takes place in the nucleus of the cell.

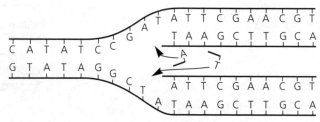

**Figure 28** Semi-conservative replication

- The enzyme DNA helicase causes the double helix to unwind.
- The DNA molecule 'unzips'. This happens because the DNA helicase causes the hydrogen bonds between the two polynucleotide strands to break.
- New DNA nucleotides join on to the exposed template strand of DNA by complementary base-pairing. A always pairs with T and C always pairs with G.
- DNA polymerase joins the new nucleotides together by a condensation reaction.

**Knowledge check 14**

Give two similarities between a DNA nucleotide and an RNA nucleotide, and two differences.

**Knowledge check 15**

How does complementary base-pairing ensure that the DNA molecule is copied accurately?

## Summary

- In all living cells, DNA holds genetic information and RNA transfers genetic information from DNA to the ribosomes.
- Both DNA and RNA are polymers of nucleotides. Each nucleotide is formed from a pentose, a nitrogen-containing organic base and a phosphate group.
- The components of a DNA nucleotide are deoxyribose, a phosphate group and one of the organic bases adenine, cytosine, guanine or thymine.
- The components of an RNA nucleotide are ribose, a phosphate group and one of the organic bases adenine, cytosine, guanine or uracil.
- A DNA molecule is a double helix with two polynucleotide chains held together by hydrogen bonds between specific complementary base pairs.
- An RNA molecule is a relatively short polynucleotide chain.
- Before cells divide, DNA copies itself accurately by a semi-conservative mechanism.

# ATP

Figure 29 shows a molecule of ATP (adenosine triphosphate).

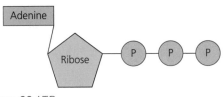

**Figure 29** ATP

You will see that it contains:

- a molecule of adenine (which is one of the nitrogen-containing organic bases found in DNA and RNA)
- a molecule of ribose (the five-carbon sugar found in RNA)
- three phosphate groups

The third phosphate group can be hydrolysed from the rest of the molecule, releasing energy that can be used for energy-requiring processes in the cell. We can say that the hydrolysis of ATP is **coupled** to energy-requiring processes in cells. When this third phosphate group is split off, it leaves adenosine diphosphate (ADP) and an inorganic phosphate (which biochemists write as Pi). This reaction is catalysed by the enzyme **ATP hydrolase**.

The inorganic phosphate released can also be used to add a phosphate group to another molecule (or to **phosphorylate** another molecule). Adding a phosphate group can make a molecule more reactive.

ADP and Pi can be joined together again, by a condensation reaction, to make ATP (Figure 30). This requires an input of energy.

The energy required to synthesise ATP from ADP and Pi can come from cellular respiration or photosynthesis. The reaction is catalysed by the enzyme **ATP synthase**.

**Knowledge check 16**

In what ways is a molecule of ATP
**a** similar to and
**b** different from an RNA nucleotide?

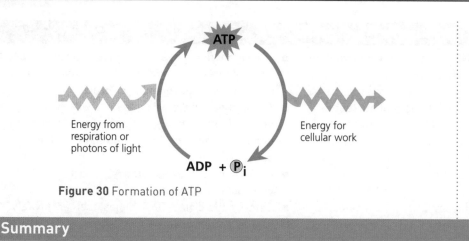

**Figure 30** Formation of ATP

## Summary

- A single molecule of adenosine triphosphate (ATP) is a nucleotide derivative and is formed from a molecule of ribose, a molecule of adenine and three phosphate groups.

- Hydrolysis of ATP to adenosine diphosphate (ADP) and an inorganic phosphate group (Pi) is catalysed by the enzyme ATP hydrolase.
- The hydrolysis of ATP can be coupled to energy-requiring reactions within cells.

# Water

Water makes up about 70% of living organisms and is essential for life on Earth. The main reason for this is that water has some unusual properties. These are because of the structure of its molecules. You can see the structure of a molecule of water in Figure 31. You can see that there is a small negative charge on the oxygen atom, and a small positive charge on each of the hydrogen atoms.

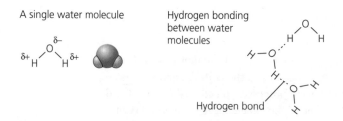

**Figure 31** Structure of a water molecule

This means that hydrogen bonds form between the negatively charged oxygen of one water molecule and one of the positively charged hydrogen atoms on another water molecule.

## The importance of water

- Water is important as a **metabolite** in hydrolysis and condensation reactions. You have already seen that many biological molecules are formed when smaller molecules join together by condensation reactions in which a molecule of water is removed. Hydrolysis reactions, when bonds are broken by the addition of a molecule of water, are required to break these larger molecules down into their components.
- Water is an excellent **solvent**. Any fairly small molecule that carries a charge can dissolve in water, because the negative charge on the oxygen atom of a

water molecule is attracted to positive charges on the solvent molecules, and the positively charged hydrogen atoms of the water molecule are attracted to negative charges on the solvent molecule. This means that water can transport substances within living organisms. For example, blood plasma is mainly water, and this transports substances such as glucose, amino acids and ions around the body of a mammal.

■ Water has a relatively high heat capacity because of the hydrogen bonding, which makes water act as a **thermal buffer**. This means it is slow to heat up and cool down. As already stated, organisms consist mainly of water. Therefore they warm up and cool down relatively slowly, so they are not subjected to sudden changes in temperature. Also, organisms that live in water are not subjected to sudden changes in temperature.

■ Water has a relatively high **latent heat of evaporation**. In other words, it takes more energy for water to form a vapour than would be expected for such a small molecule. This is because energy is needed to break the hydrogen bonds between water molecules when it vaporises. As a result, water is very effective for cooling organisms down. For example, in humans, a small amount of sweat (which is mainly water) evaporating from the skin can have a significant cooling effect.

■ There is strong **cohesion** between water molecules because of the hydrogen bonding. This means that water molecules 'stick' together. For example, when columns of water travel up the xylem in flowering plants, they are being 'pulled' by evaporation from the leaves. These strong cohesive forces also mean that water has a **high surface tension**, creating a 'skin' on top of the surface. This allows insects like pond skaters to move on the surface of ponds, or mosquito larvae to suspend themselves below the water surface.

# Inorganic ions

Inorganic ions are present in the cytoplasm of cells and dissolved in the body fluids of organisms. They can be present in high or low concentrations.

Pure water has a neutral pH, i.e. a pH of 7. pH is a measurement of the hydrogen ion ($H^+$) concentration in a substance. In water, a small number of the molecules split up. This produces a hydrogen ion ($H^+$) and a hydroxide ion ($OH^-$). The hydrogen ions then combine with water molecules to form hydronium ions ($H_3O^+$) but, to make things easier, these are simply regarded as hydrogen ions ($H^+$). In pure water, there is an equal number of hydrogen ions and hydroxide ions, so the pH is neutral, pH 7.

Acids are substances that donate hydrogen ions, so when an acid is dissolved in water there are more hydrogen ions present than hydroxide ions. This means the solution is acidic and has a pH lower than 7. On the other hand, a base is a substance that accepts hydrogen ions. When a base is dissolved in water it takes up hydrogen ions, so now the concentration of hydrogen ions is lower than the concentration of hydroxide ions. This results in an alkaline solution with a pH higher than 7.

Other inorganic ions are important in biology. Examples include iron ions, which form part of the protein haemoglobin and enable it to carry oxygen effectively. Sodium ions are involved in the co-transport of glucose and amino acids from the gut lumen into the epithelial cells of the small intestine. Phosphate ions are important in the structure of DNA and RNA nucleotides, as well as in ATP.

**Knowledge check 17**

Explain how phosphate ions are soluble in water.

**Exam tip**

The **pH** scale used to measure acidity and alkalinity is a **logarithmic** scale. This is because a strongly acidic solution can have 100 000 000 000 000 times more hydrogen ions than a strongly basic solution, and vice versa. In order to deal with these large numbers more easily, scientists use a logarithmic scale, the pH scale. Each one-unit change in the pH scale corresponds to a ten-fold change in hydrogen ion concentration. The pH scale ranges from 0 to 14. Using a logarithmic scale saves you having to write down all the zeros.

## Summary

- Water is a metabolite in many metabolic reactions, including condensation and hydrolysis reactions.
- It is an important solvent in which metabolic reactions occur.
- It has a relatively high heat capacity, buffering changes in temperature.
- It has a relatively large latent heat of vaporisation, providing a cooling effect with little loss of water through evaporation.
- It has strong cohesion between its molecules.
- pH measures the concentration of hydrogen ions in a solution.
- Iron ions are a component of haemoglobin.
- Sodium ions are involved in the co-transport of glucose and amino acids.
- Phosphate ions are components of DNA and ATP.

# ■ Cells

## Cell structure

### Structure of eukaryotic cells

The structure of an animal cell is shown in Figure 32, and the structure of a plant cell is shown in Figure 33.

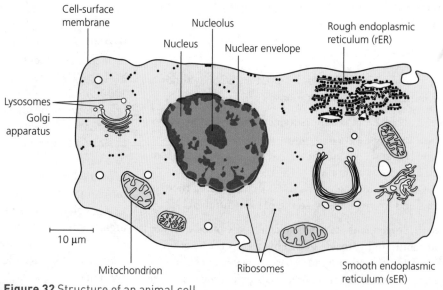

**Figure 32** Structure of an animal cell

You can see that these cells contain many smaller components, called **organelles**. The function of these organelles are outlined in Table 4. Plant and animal cells are **eukaryotic** cells. This is because they contain a membrane-bound nucleus and membrane-bound organelles such as mitochondria.

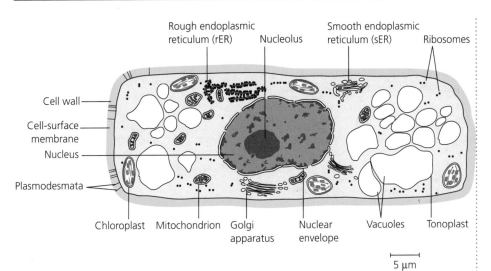

**Figure 33** Structure of a plant cell

| Organelle | Structure | Function |
|---|---|---|
| Ribosomes | Very small organelle, made of RNA and protein<br>They are not surrounded by a membrane | Carry out protein synthesis |
| Rough endoplasmic reticulum | Layers of membranes that form flattened, interconnected tubes through the cytoplasm<br>The outer face is covered in ribosomes | Synthesise proteins and transport them to the Golgi apparatus |
| Golgi apparatus | Stacks of flattened, membrane-bound sacs<br>Vesicles are constantly pinched off the ends of these sacs | Modify proteins made at the rER, for example, adding carbohydrate to form glycoproteins<br>These are secreted from the cell in Golgi vesicles, pinched off from the cisternae |
| Mitochondria | Bean-shaped organelles that have a double membrane around them<br>The inner membrane is folded into cristae, and the matrix in the middle contains enzymes | Aerobic respiration |
| Lysosomes | Membrane-bound vesicles containing hydrolytic enzymes (lysozymes) produced by the Golgi apparatus | Digest unwanted materials in the cell, for example, worn-out mitochondria |
| Smooth endoplasmic reticulum | Similar to rER but the cavities are more tubular and the membranes do not have ribosomes | Synthesis of lipids |
| Cell surface membrane | Made of phospholipids and proteins<br>Surrounds the cell | Controls the passage of substances into and out of the cell |
| Nucleus | This is the largest organelle in the cell<br>It is surrounded by a double membrane called the nuclear envelope, which has many holes in it called nuclear pores | Contains DNA, which holds the genetic information necessary to control the cell<br>The DNA is in the form of linear chromosomes, made of DNA wound around proteins |
| Nucleolus | A dark, granular area inside the nucleus that contains DNA and proteins | Synthesises ribosomes |

| Organelle | Structure | Function |
|---|---|---|
| Chloroplasts (found in plants and algae only) | Surrounded by a double membrane Contain stacks of membranes that contain chlorophyll and other pigments, surrounded by the liquid stroma, which contains enzymes | Carry out photosynthesis |
| Vacuole (in plants) | A space surrounded by a membrane called the tonoplast and containing cell sap | Keeps cells turgid |
| Cell wall (in plants, algae and fungi) | In plants, the cell wall is made of cellulose fibres and surrounds the cell surface membrane Algal cell walls are typically made of glycoproteins and polysaccharides, and fungal cell walls are usually made of the carbohydrate chitin | The cell wall surrounds the cell membrane and stops the cell bursting when it takes in water by osmosis |

**Table 4** The functions of organelles in a eukaryotic cell

Figure 34 shows the relationship between the rough endoplasmic reticulum, the Golgi apparatus and vesicles in producing proteins (such as enzymes) or glycoproteins (such as mucus) that are secreted outside the cell.

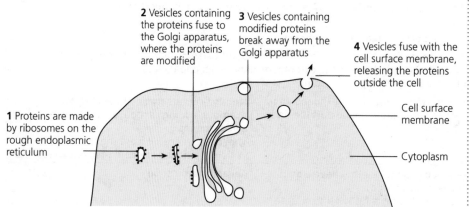

**2** Vesicles containing the proteins fuse to the Golgi apparatus, where the proteins are modified

**3** Vesicles containing modified proteins break away from the Golgi apparatus

**4** Vesicles fuse with the cell surface membrane, releasing the proteins outside the cell

**1** Proteins are made by ribosomes on the rough endoplasmic reticulum

Cell surface membrane

Cytoplasm

**Figure 34** The organelles involved in making proteins that are secreted outside the cell

### Tissues, organs and systems

Many organisms are multicellular. They contain many different kinds of cell, each specialised to perform a specific function.

- A **tissue** is a group of similar cells that carries out a specific function, for example, blood tissue in mammals or xylem tissue in flowering plants.
- An **organ** is made of several different tissues, which work together to carry out a specific function. For example, a leaf contains xylem, phloem, palisade mesophyll and spongy mesophyll cells. The stomach contains many different tissues such as smooth muscle tissue, blood tissue, glandular tissue and connective tissue.
- Organs work together to form a **system**. For example, the heart, veins, arteries and capillaries all work together in the circulatory system to transport substances around the body.

**Knowledge check 18**

Explain why:

**a** cells in the pancreas that produce enzymes contain a lot of rough endoplasmic reticulum

**b** bird wing muscle cells contain more mitochondria than most of their other cell types, for example, blood cells

**Knowledge check 19**

Explain why blood is classed as a tissue.

## Summary

- Eukaryotic cells, such as plant and animal cells, are surrounded by a cell-surface membrane that controls what enters and leaves the cell. They also contain a number of specific organelles with specialist functions.

- In complex multicellular organisms, eukaryotic cells become specialised for specific functions. Specialised cells are organised into tissues, tissues into organs and organs into systems.

# Structure of prokaryotic cells and of viruses

## Prokaryotic cells

Prokaryotic cells are much smaller than eukaryotic cells. Prokaryotic cells are bacteria. Figure 35 shows the structure of a prokaryotic cell.

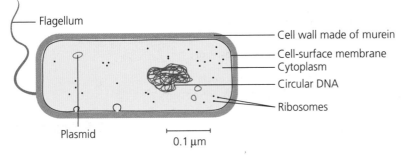

**Figure 35** Structure of a prokaryotic cell

Prokaryotic cells:

- do not contain a nucleus or a nuclear envelope
- have a single circular piece of DNA that is not complexed with proteins; this carries the cell's genetic information
- do not contain any membrane-bound organelles, such as mitochondria, chloroplasts, endoplasmic reticulum, lysosomes or Golgi apparatus
- have a cell wall surrounding the cell-surface membrane, but it is made of a glycoprotein called murein
- have ribosomes that synthesise proteins, but these are smaller than the ribosomes in eukaryotic cells

In addition, some prokaryotic cells:

- contain **plasmids**, which are small, circular pieces of DNA that carry genes additional to those in the main piece of circular DNA
- have one or more **flagella**, that enable the cell to move
- have a slimy capsule surrounding the cell, which stops the cell drying out, helps to protect the cell from being engulfed by white blood cells, and stores toxins

## Viruses

Viruses are much smaller than prokaryotic cells. They are **acellular**, which means they are not made of cells, and they are non-living. The structure of a virus is shown in Figure 36.

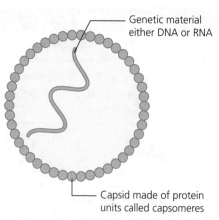

Genetic material
either DNA or RNA

Capsid made of protein
units called capsomeres

**Figure 36** Structure of a virus

You will see that viruses contain genetic material. This can be DNA or RNA but never both. Surrounding the genetic material is a protein coat, called a **capsid**, made of many protein units. Some viruses have a membrane around the outside.

Viruses do not show any signs of life. They do not metabolise, grow or respire, or any of the other characteristics that are associated with living things. They can only reproduce when they are inside a living cell. On the outside of the virus is an **attachment protein**, which fits into a protein in the membrane of a living host cell. This enables the virus to enter the cell. Once inside the cell, their genetic material 'hijacks' the cell's organelles. The organelles start to make new viral proteins and the cell starts making many new viruses. These viruses are then released from the host cell, destroying the host cell in the process.

**Knowledge check 20**

Which enzymes in the host cell will be used to copy the viral DNA?

## Summary

- Prokaryotic cells are much smaller than eukaryotic cells. They also differ from eukaryotic cells in having:
  - no membrane-bound organelles, such as nuclei or mitochondria
  - cell walls made of murein

- DNA that is circular and not complexed with protein
- Viruses are acellular and non-living. Virus particles contain genetic material, surrounded by a capsid. They have an attachment protein on their surface to help them enter host cells.

# Methods of studying cells

## Microscopes

Many structures in biology, such as cells and the structures within them, are too small to be seen by the naked eye. An **optical microscope** can be used to see the structure of some of these objects.

An optical microscope works by passing a beam of light through a specimen, magnifying and focusing the image using lenses. This means that the specimen needs to be quite thin. Sometimes a stain is added, to make part of the specimen a different colour from the rest, so that the structures can be seen more clearly. However, there is a limit to what can be seen with an optical microscope.

**Magnification** is making things larger. **Resolution** is the ability to see two objects that are close together as separate objects. There is a limit to the resolution of an optical microscope, as it cannot resolve two objects that are closer together than half the wavelength of the light being used in the microscope. Although you can see eukaryotic cells from animals and plants using an optical microscope, it is no use for studying the tiny organelles within the cells or the structure of prokaryotic cells.

The **transmission electron microscope** uses a beam of electrons, which has a much smaller wavelength than light. Therefore its resolution is much greater than that of an optical microscope.

This kind of microscope allows scientists to study the detailed structure of cells — their **ultrastructure**. However, the specimen needs lengthy preparation, staining with heavy metals and cutting into extremely thin sections. This means that a highly skilled technician is needed.

Inside the microscope there must be a vacuum so that the electron beam can pass through without being scattered. Therefore you cannot study a living organism using an electron microscope. Also, you cannot view the image directly, as our eyes are not sensitive to electron beams. The image can only be seen on a computer screen. Also, the image can only be seen in black and white, not colour. Many electron micrographs that you see in books have colours in them, but these have been added using a computer. They are not the real colours of the structures.

The **scanning electron microscope** works rather like the transmission electron microscope, except that the electron beam bounces off the surface of the object. This enables you to see the shape of structures such as viruses.

One problem that is faced when using microscopes is that the specimen has to be carefully prepared. This can mean adding chemicals to stabilise its structure, embedding it in wax or resin, adding stains and slicing it thinly before putting it onto a slide or grid to view under the microscope.

In doing this, the structure of the specimen can be altered. This means that a structure might be seen in the image that was not there in the original living specimen. We call these structures **artefacts**. For example, for many years, infoldings of the cell-surface membrane called mesosomes were seen in bacterial cells prepared for electron microscopy. However, when a new system of preparing specimens called freeze-drying was developed, mesosomes were no longer seen. We now know that these infoldings were artefacts caused by the way the specimen was prepared.

### Calculating the size of objects

We need to use different units to measure the size of objects, depending on their size.
- 1 millimetre (mm) is $10^{-3}$ m (i.e. there are 1000 mm in 1 m).
- 1 micrometre (μm) is $10^{-6}$ m (ie there are 1000 μm in 1 mm, or 1 000 000 μm in 1 m).
- A nanometre (nm) is $10^{-9}$ m (ie there are 1 000 000 nm in 1 mm, or 1 000 000 000 nm in 1 m).

You can work out the magnification of an image using the formula:

$$\text{magnification} = \frac{\text{measured size}}{\text{actual size}}$$

**Exam tip**

Remember that the resolution of an electron microscope is higher than that of an optical microscope, because electron beams have a smaller wavelength than light. Talking about higher magnification will not get you marks.

**Worked examples**

1 Calculate the magnification of the animal cell in Figure 32 on page 24.

This diagram has a scale bar, so we measure the length of the scale bar in mm. This is 12 mm.

We convert mm to μm because we must use the same units as the actual length, and the actual length is 10 μm. So the measured length = 12 × 1000 = 12 000 μm.

$$\text{magnification} = \frac{\text{measured size}}{\text{actual size}} = \frac{12\,000}{10} = 1200$$

2 Figure 37 shows a mitochondrion drawn from an electron micrograph. Calculate the actual length of the mitochondrion.

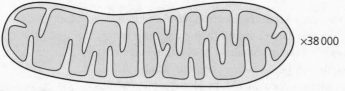

×38 000

**Figure 37**

This time you are given the magnification and you can measure the size of the image, but you need to find the actual size. This can be done by rearranging the equation:

$$\text{actual size} = \frac{\text{measured size}}{\text{magnification}}$$

Measure the length of the image in mm (in this case 80 mm).

Convert mm to micrometres by multiplying by 1000 = 80 000.

$$\text{actual size} = \frac{\text{measured size}}{\text{magnification}} = \frac{80\,000}{38\,000} = 2.11\,\mu\text{m}$$

## Cell fractionation and ultracentrifugation

Scientists studying cells sometimes find it useful to separate out the different organelles of a cell, so that they can study them. This works because the different cell organelles have different **densities**.

It is important that fresh tissue is used so that the organelles are still functional and undamaged.

- Cells are broken open using a homogeniser.
- The mixture produced is spun at high speed in a centrifuge. This forces the densest material to the bottom of the tube where it forms a pellet.
- The liquid above the pellet is called the supernatant. This is spun again in a centrifuge at higher speed. This time the organelles that are next in density are present in the pellet.

**Knowledge check 21**

Calculate the magnification of the plant cell in Figure 33 and the bacterial cell in Figure 35.

**Exam tip**

Remember that cell fractionation separates out the densest organelles first, not the largest or heaviest. Words are really important when answering questions on this topic.

Figure 38 shows the processes in cell fractionation and ultracentrifugation.

(1)
Chop up fresh liver tissue in ice-cold isotonic buffer solution

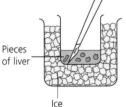

Pieces of liver

Ice

(2)
Put the chopped tissue into a blender or homogeniser, which breaks open the cells

(3)
Filter the mixture to remove the debris

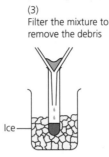

Ice

(4)
Pour the mixture into tubes and spin very quickly in a centrifuge. The denser parts of the mixture get spun to the bottom of the tube where they form a pellet, called the 'sediment'.

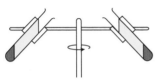

(5)
The liquid layer on top (the supernatant) is poured into a fresh tube, leaving the sediment behind (this contains the nuclei)

(6)
The supernatant can then be spun again at a faster speed to produce a sediment containing mitochondria

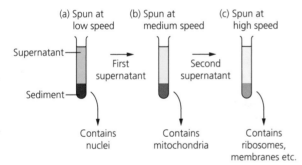

(a) Spun at low speed    (b) Spun at medium speed    (c) Spun at high speed

Supernatant

First supernatant    Second supernatant

Sediment

Contains nuclei    Contains mitochondria    Contains ribosomes, membranes etc.

**Figure 38** Cell fractionation and ultracentrifugation

**Exam tip**

Make sure that you understand that the isotonic solution stops the organelles bursting by osmosis. Many students say that it stops the cells from bursting — but of course they are being deliberately burst by using the homogeniser, so you won't get any marks for this.

In stage 1 in Figure 38, the buffer solution keeps the pH of the cell contents constant. Any change in pH could denature the enzymes in the mixture, which means that any organelles isolated would not be very useful. Everything is kept ice cold, so that the enzymes do not denature, but also so they are not active. If the enzymes were active they might digest the organelles. The buffer solution is isotonic so that the organelles do not take in water by osmosis and burst.

**Knowledge check 22**

How would you use this technique to obtain a sample of chloroplasts?

### Summary

- The structure of cells can only be understood by using microscopes.
- The resolution of an optical microscope is limited by the wavelength of light. To study very small structures, such as cell organelles, better resolution is needed.
- Resolution is the ability to see two objects close together as separate objects. Magnification is making something appear larger.
- Electron microscopes use a beam of electrons, which has a much smaller wavelength than light, so they have much better resolution.
- The real size of a magnified image, or the magnification of an image, can be calculated using the formula:

$$\text{magnification} = \frac{\text{size of image}}{\text{size of real object}}$$

- Cell fractionation and ultracentrifugation are used to separate cell components.

# All cells arise from other cells

## Mitosis

Multicellular organisms start life as a single cell. This cell divides repeatedly to produce all the cells in the adult organism. The kind of cell division required for growth that results in identical copies of the cell being made is called **mitosis**. During growth, cells go through a **cell cycle**, which involves growth phases as well as mitosis. You can see this in Figure 39.

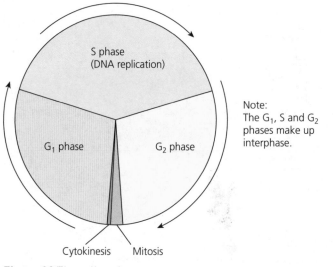

Note:
The $G_1$, S and $G_2$ phases make up interphase.

**Figure 39** The cell cycle

Most of the time a cell is in **interphase**. Interphase is when the cell is carrying out its normal activities, such as producing more cytoplasm and organelles, respiration and protein synthesis. The $G_1$, S and $G_2$ phases are all parts of interphase. During the S phase of the cell cycle the DNA replicates ready for mitosis, so there are two identical copies of each DNA molecule in the nucleus. Each DNA molecule forms one chromosome. Therefore, after DNA replication, each chromosome consists of two identical DNA molecules called **chromatids** that are joined together at a point called the centromere. This is shown in Figure 40.

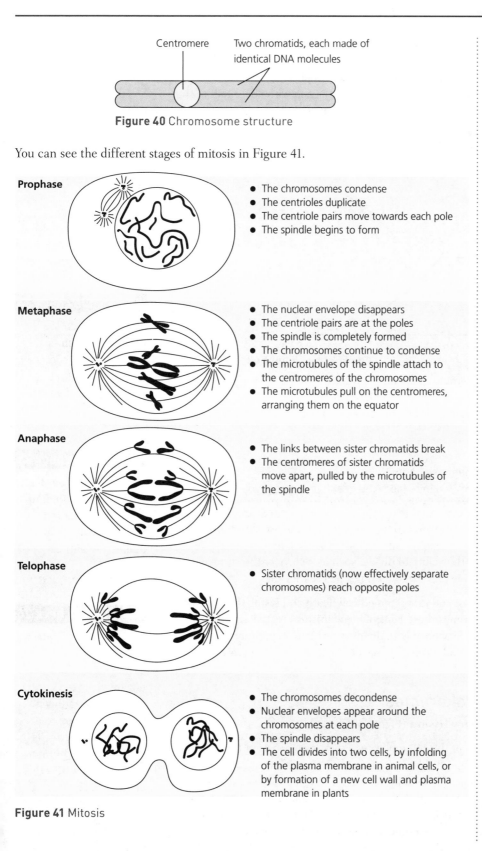

Centromere    Two chromatids, each made of identical DNA molecules

**Figure 40** Chromosome structure

You can see the different stages of mitosis in Figure 41.

**Prophase**
- The chromosomes condense
- The centrioles duplicate
- The centriole pairs move towards each pole
- The spindle begins to form

**Metaphase**
- The nuclear envelope disappears
- The centriole pairs are at the poles
- The spindle is completely formed
- The chromosomes continue to condense
- The microtubules of the spindle attach to the centromeres of the chromosomes
- The microtubules pull on the centromeres, arranging them on the equator

**Anaphase**
- The links between sister chromatids break
- The centromeres of sister chromatids move apart, pulled by the microtubules of the spindle

**Telophase**
- Sister chromatids (now effectively separate chromosomes) reach opposite poles

**Cytokinesis**
- The chromosomes decondense
- Nuclear envelopes appear around the chromosomes at each pole
- The spindle disappears
- The cell divides into two cells, by infolding of the plasma membrane in animal cells, or by formation of a new cell wall and plasma membrane in plants

**Figure 41** Mitosis

**Exam tip**

Make sure you understand the difference between a chromatid and a chromosome. During interphase the chromosomes cannot be seen in the nucleus, although the DNA is there. At the start of mitosis, in prophase, the chromosomes condense out and appear. Each chromosome is already composed of two chromatids because the DNA replicated during interphase.

**Exam tip**

You can remember the stages of mitosis by the nonsense word 'IPMAT'.

During mitosis, the two chromatids split apart with each chromatid going to a different end of the cell. A new nuclear envelope forms around each group of chromatids (now called chromosomes). After this, the cytoplasm of the cell usually splits in two, forming two genetically identical new cells, each with about half of the organelles of the parent cell. This stage is called **cytokinesis**.

## Knowledge check 23

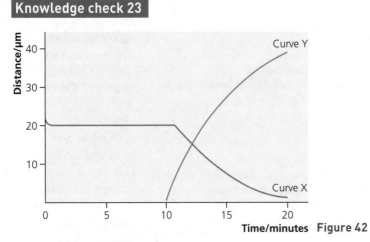

**Time/minutes** **Figure 42**

Figure 42 shows the movement of chromosomes during mitosis. Curve X shows the mean distance between the centromeres of the chromosomes and the nearest pole of the spindle.

**a** What does curve Y represent?
**b i** At what time did anaphase begin?
  **ii** Explain one piece of evidence from the graph to support your answer.

When cells in a multicellular organism differentiate, they often lose the ability to divide by mitosis. Most cells in the adult human have lost the ability to divide by mitosis. Some cells do retain the ability to divide, such as cells near the surface of the skin, but even these cells divide only 20 to 50 times before they die.

Mitosis is a controlled process, and programmed cell death and losing the ability to divide are part of this. Sometimes these control mechanisms break down and this means that the resulting cell may be able to divide repeatedly in an uncontrolled way. These are **cancer** cells, and given a supply of nutrients they can grow into a **tumour**.

## Binary fission in prokaryotes

Bacterial cells divide by **binary fission** (Figure 43).

First of all, the circular DNA of the cell replicates. These two copies of the DNA are attached to the cell's membrane. If there are plasmids in the cell, these also replicate. The cell then grows, increasing the distance between the two pieces of circular DNA. A new cell wall starts to grow between the pieces of DNA, and more cell membrane is synthesised, so that the cell splits in two. Each daughter cell has one copy of the main circular piece of DNA, and a variable number of copies of the plasmids.

### Knowledge check 24

A substance called colchicine inhibits spindle formation. Explain the effect that this will have on mitosis.

### Knowledge check 25

Cytoxan is a drug that stops DNA unwinding prior to replication. Explain how this is useful in treating cancer.

### Knowledge check 26

Ciprofloxacin is a drug that inhibits the enzyme DNA gyrase. This enzyme relaxes tightly wound DNA, allowing DNA replication to occur. DNA gyrase is not present in human cells. Explain how this drug might be useful in treating bacterial diseases.

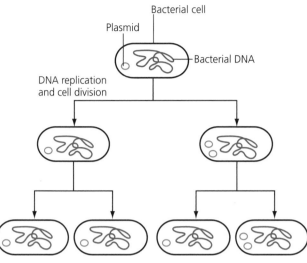

**Figure 43** Binary fission

# Viral replication

- The virus attaches to the cell membrane of its host cell. The viral attachment protein fits into a protein on the cell membrane of the host cell.
- The virus is taken into the cell and the capsid is removed.
- If the viral genetic material is RNA, it moves to the ribosomes and the ribosomes start to produce enzymes to synthesise new virus particles. If the viral genetic material is DNA, this acts as the genetic code for RNA to be produced that codes for the production of viral enzymes and proteins.
- New viruses are assembled by the host cell. Sometimes they leave the cell by 'budding', which means they are coated with some membrane from the host cell as they leave. Sometimes they burst out of the host cell by rupturing the cell membrane.

Two possible ways in which a virus replicates are shown in Figure 44.

**Knowledge check 27**

Use the information about viral replication to explain why viruses are described as non-living.

## Required practical 2

### Root tip squash

Plants have groups of dividing cells in certain places. One of these places is at the root tip, just behind the root cap. It is possible to see many cells undergoing mitosis if you make a microscope slide of this region of the root. However, to view this tissue under an optical microscope, the sample needs to be very thin, and a stain needs to be added so that the chromosomes show up.

To make a root tip squash, cut 5mm off the end of a young growing root. Put this into a glass dish containing dilute hydrochloric acid and acetic orcein stain. Then warm this carefully. The hydrochloric acid separates the cells, so the tissue becomes softer, and the acetic orcein stains the DNA in the chromosomes red.

Next put the root tip into the middle of a microscope slide. Add two drops of acetic orcein stain to it, and carefully lower a cover slip on top. Fold a piece of filter paper and place it carefully on top, then push firmly. This squashes the tissue and flattens it into a thin layer only one cell thick.

After this, view the slide under an optical microscope and identify which stage of mitosis the cells are in.

1 How could you use this investigation to estimate how long each stage of mitosis lasts?

Turn to page 91 for the answer.

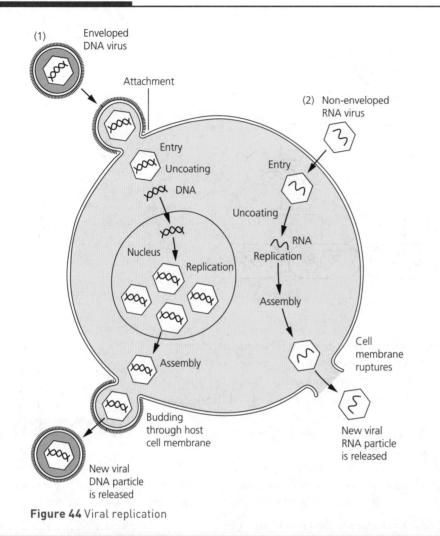

**Figure 44** Viral replication

## Summary

- DNA replication occurs during the interphase of the cell cycle when chromosomes cannot be seen.
- Mitosis is the part of the cell cycle in which a eukaryotic cell divides to produce two daughter cells, each with the identical copies of DNA produced by the parent cell during DNA replication.
- During prophase, the chromosomes condense out. In metaphase, the chromosomes attach to the equator of the spindle. During anaphase, the spindle fibres separate the sister chromatids, and in telophase a new nuclear membrane forms around each identical set of chromatids (now daughter chromosomes).

- Division of the cytoplasm (cytokinesis) usually occurs, producing two new cells.
- Mitosis is a controlled process. Uncontrolled cell division can lead to the formation of tumours and of cancers. Many cancer treatments are directed at controlling the rate of cell division.
- Binary fission in prokaryotic cells involves:
  - replication of the circular DNA and of plasmids
  - division of the cytoplasm to produce two daughter cells, each with a single copy of the circular DNA and a variable number of copies of plasmids
- Being non-living, viruses do not undergo cell division. Following injection of their nucleic acid, the infected host cell replicates the virus particles.

# Transport across cell membranes

## The structure of cell membranes

A cell membrane is made of a double layer of **phospholipid** molecules. You will remember that the 'head' of a phospholipid molecule carries a charge because of its phosphate group, so it is **hydrophilic**. The fatty acid 'tail' is **hydrophobic**. Therefore, in water phospholipid molecules arrange themselves with the 'heads' towards water and the 'tails' away from water, forming a **bilayer** (Figure 45).

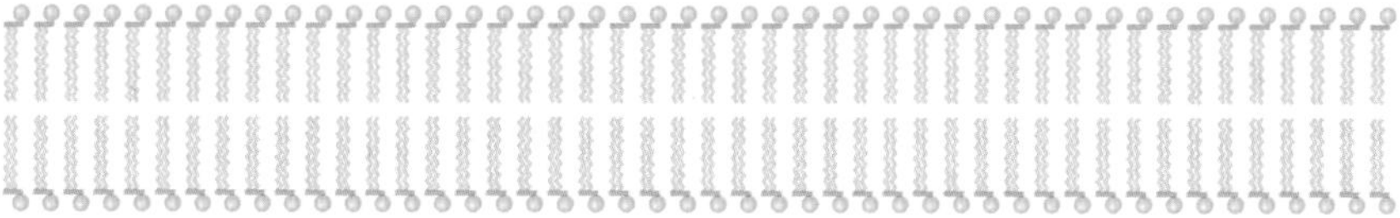

**Figure 45** Phospholipid bilayer

**Knowledge check 28**

Explain why the phosphate group makes the 'head' of a phospholipid hydrophilic.

**Protein** molecules float in the phospholipid bilayer. Some of these are **extrinsic** proteins, which float in just one half of the bilayer. Others are **intrinsic** and pass through the whole bilayer. **Cholesterol** molecules are found among the phospholipids. These stabilise the membrane by restricting the movement of other molecules within the membrane. Many of the proteins and phospholipids have short carbohydrate chains attached to them, on the outside surface of the membrane. They are known as **glycoproteins** and **glycolipids**. The structure of the cell membrane is shown in Figure 46.

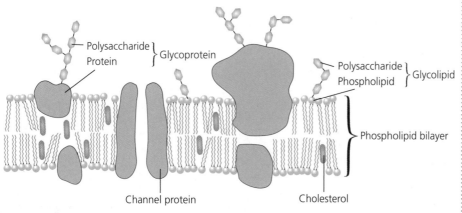

**Figure 46** The cell membrane

This structure is described as **fluid mosaic** because:

- the phospholipids move around within their layer, making the membrane fluid
- the proteins are scattered among the phospholipids, creating a mosaic

Proteins within the cell membrane can function as:

- channels
- carriers
- antigens
- receptors
- enzymes

All the membranes in a cell have this fluid mosaic structure. This includes the membranes of the chloroplast, mitochondria, endoplasmic reticulum and lysosomes.

## Movement across cell membranes

### Simple diffusion

Diffusion is the movement of molecules from a region of high concentration to a region of lower concentration down a concentration gradient.

Some molecules that are small and/or lipid soluble are able to diffuse through the phospholipid bilayer. Oxygen, for example, is a small molecule and it is not charged, so it is able to pass freely through the phospholipid bilayer. It diffuses from outside the cell where it is in higher concentration, across the membrane into the cell, down a concentration gradient. Simple diffusion is outlined in Figure 47.

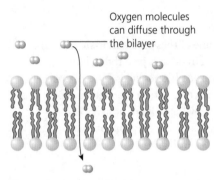

**Figure 47** Simple diffusion

Simple diffusion is described as a **passive** process because it does not require additional energy.

### Facilitated diffusion

Facilitated means 'helped'. Ions or molecules that carry a charge cannot pass through the phospholipid bilayer. Also, large molecules cannot pass between the phospholipids. These ions and molecules can only pass through the membrane with the help of proteins, which is why this kind of diffusion is described as 'facilitated'.

Some of the proteins are **channel proteins**. These have a hydrophilic channel through them that allows specific molecules through. Some of the other proteins are **carrier proteins**. These are also specific to certain kinds of ion or molecule. The protein carrier changes shape when the molecule binds, and releases it on the other side of the membrane. Facilitated diffusion is outlined in Figure 48.

Facilitated diffusion, like simple diffusion, is a passive process because no additional energy is needed.

**Exam tip**

Make sure you describe diffusion and facilitated diffusion as movement down a concentration gradient, not 'along' it.

**Knowledge check 29**

Use your knowledge of protein structure to explain how carrier and channel proteins can be specific to just one kind of molecule.

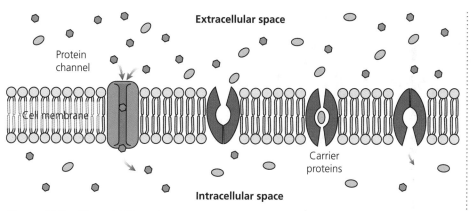

**Figure 48** Facilitated diffusion

### Knowledge check 30

Figure 49 shows the rate of diffusion of two different molecules into a cell. One molecule is entering the cell by simple diffusion and one by facilitated diffusion. Which is which? Give a reason for your answer.

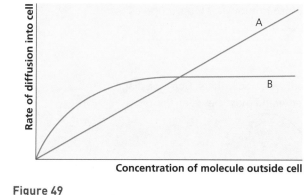

**Figure 49**

## Osmosis

Water molecules are small, and although they carry charges they can slip between the phospholipid molecules in the bilayer. They enter cells by a special kind of diffusion called **osmosis**.

The more water molecules that are present in a solution, the higher the **water potential** of the solution. Pure water has the highest possible concentration of water molecules, so it has the highest possible water potential of 0. All solutions have a lower water potential than this. Concentrated solutions, with a lot of solute, have a low concentration of water so they have a low water potential, i.e. a negative water potential. In osmosis, water diffuses from a region of high water potential to a region of lower water potential down a water potential gradient.

Osmosis is outlined in Figure 50.

Osmosis is also a passive process because it does not require additional energy.

### Exam tip

Always refer to water potentials and not water concentrations. But it helps you to remember that a solution with a high water concentration has a high water potential.

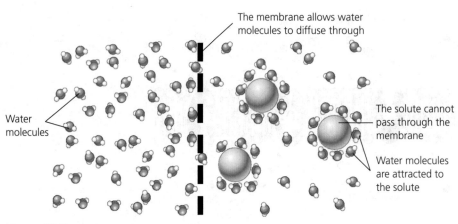

Figure 50 Osmosis

**Exam tip**

Water potentials are measured in pressure units and the numbers are negative. Remember that water moves from a high (less negative) to a lower (more negative) water potential. It may help you to think of temperatures. Heat is transferred from a higher temperature to a lower temperature, so heat would move from an object at –1°C to an object at –5°C.

**Knowledge check 31**

a The three cells in Figure 51 have different water potentials. Copy the diagram and add arrows to show which way water would move between them.

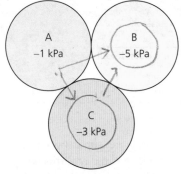

Figure 51

b Which of your arrows represents the fastest rate of water movement?

**Knowledge check 32**

a If you put a red blood cell into water, it will burst. Explain why.

b If you put a plant cell into water, it will not burst. Explain why.

## Active transport

Sometimes cells take up substances against their concentration gradient. In this case, a process called **active transport** is used (Figure 52). Active transport, like facilitated diffusion, uses carrier proteins that are specific to the molecule being carried. However, active transport uses additional energy because the molecule is being moved against its concentration gradient. This is why it is called an active process. The energy for active transport comes from the hydrolysis of ATP:

ATP → ADP + energy

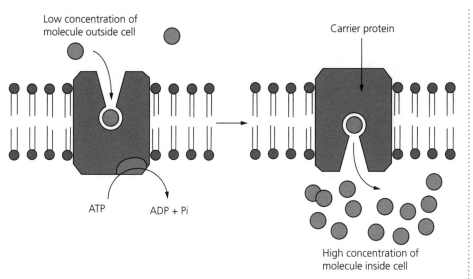

Figure 52 Active transport

**Knowledge check 33**

Figure 53 shows the concentration of two ions inside and outside a cell.

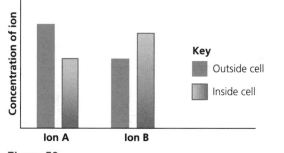

Figure 53

How do ions A and B enter the cell? Give reasons for your answer.

## Co-transport

Figure 54 shows how glucose is absorbed into the epithelial cells on the villi of the small intestine.

There are three important protein carriers in this cell:

- **Protein A** actively transports sodium ions out of the cell, creating a low concentration of sodium ions inside the cell.
- **Protein B** is a **co-transport protein**. It carries glucose into the cell, alongside a sodium ion, which diffuses into the cell down its concentration gradient. It is called a co-transport protein because it has to have both a sodium ion and a glucose molecule to bring into the cell.
- **Protein C** is a carrier protein that transports glucose out of the cell into the blood capillary by facilitated diffusion.

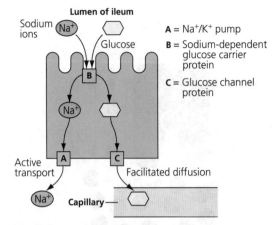

**Figure 54** Glucose absorption

## Required practicals 3 and 4

### Investigating osmosis

#### Investigation to find the water potential of potato tissue

A student set up seven Petri dishes so that each one contained a different concentration of sucrose. She made up these solutions using 1.0 M sucrose and distilled water. Table 5 shows how she did this.

| Concentration of sucrose solution in dish/M dm⁻³ | Volume of 1.0 M sucrose added/cm³ | Volume of distilled water added/cm³ |
|---|---|---|
| 0.0 | 0 | 20 |
| 0.2 | 4 | 16 |
| 0.4 | 8 | 12 |
| 0.6 | 12 | 8 |
| 0.8 | 16 | 4 |
| 1.0 | 20 | 0 |

**Table 5**

She then cut several cores of potato tissue using a cork borer, and sliced them into slices about 2 mm thick. She washed them in distilled water and gently blotted them dry using filter paper. She weighed the discs in batches of five, recorded the mass, and then put five discs in each dish. The dishes were left in a cool place overnight.

The following day she removed the discs from each dish in turn, carefully blotted them dry and re-weighed them. She recorded the mass.

The student then calculated the percentage change in mass and plotted a graph of her results. The results are shown in Figure 55.

1 How did the student calculate the percentage change in mass?
2 Why was it better to calculate the percentage change in mass, rather than just the change in mass?
3 Use your knowledge of water potential to explain why the discs placed in 1.0 M sucrose lost mass, and those in 0.1 M sucrose gained mass.

---

**Knowledge check 34**

The epithelial cell from the lining of the small intestine shown in Figure 54 has many infoldings of its cell-surface membrane on the face of the cell that borders the lumen of the gut. These are called microvilli. Suggest the advantage of this.

**Knowledge check 35**

Explain why protein A in Figure 54 is necessary for the co-transport protein to bring glucose molecules into the cell.

**Exam tip**

Make sure you know how to make up a series of dilutions, starting with a standard solution.

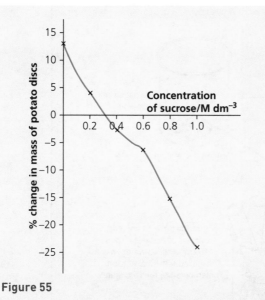

**Figure 55**

The student found the sucrose solution with the same water potential as the potato tissue by finding the concentration of sucrose where the line crosses the x-axis. This is the concentration of sucrose where the potato tissue would not gain or lose mass. In this case, it crosses the x-axis at 0.32 M.

## Investigating the effect of temperature on the permeability of beetroot cell membrane

Beetroot cells contain a red pigment, so that when the cell membrane breaks down the red pigment leaks out. The more permeable the membrane has become, the more red pigment leaks out.

A student cut several cores of beetroot using a cork borer and then cut them into cylinders 3 cm long. He stored them in a beaker of distilled water. The student then pipetted 15 cm$^3$ of distilled water into each of 10 test tubes. He set up a water bath at 70°C, dropped a cylinder of beetroot into the water bath for exactly 1 minute and then placed the cylinder into one of the test tubes of water. The beetroot cylinder was left in the tube of water at room temperature for exactly 15 minutes. After this time, the student carefully removed the beetroot cylinder from the tube using forceps. He cooled the water bath to 65°C and then carried out the same process. This was repeated until he had carried out the investigation at 25, 30, 35, 40, 45, 50, 55, 60, 65 and 70°C. See Figure 56.

The student measured the concentration of the red pigment in each tube using a colorimeter.

A colorimeter is an instrument that shines light of a particular wavelength through a solution. The more intense the colour of the solution, the more light is absorbed by the solution and therefore less light is transmitted through. The colorimeter measures how much light is absorbed and how much light is transmitted through.

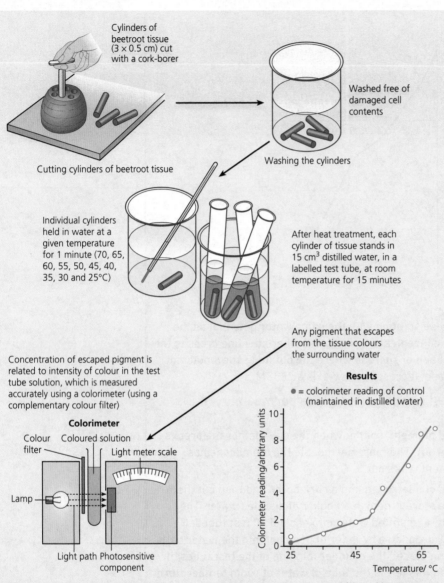

Cylinders of beetroot tissue (3 × 0.5 cm) cut with a cork-borer

Washed free of damaged cell contents

Cutting cylinders of beetroot tissue

Washing the cylinders

Individual cylinders held in water at a given temperature for 1 minute (70, 65, 60, 55, 50, 45, 40, 35, 30 and 25°C)

After heat treatment, each cylinder of tissue stands in 15 cm³ distilled water, in a labelled test tube, at room temperature for 15 minutes

Any pigment that escapes from the tissue colours the surrounding water

Concentration of escaped pigment is related to intensity of colour in the test tube solution, which is measured accurately using a colorimeter (using a complementary colour filter)

**Results**

● = colorimeter reading of control (maintained in distilled water)

**Colorimeter**

Colour filter  Coloured solution

Light meter scale

Lamp

Light path  Photosensitive component

**Figure 56** Investigating the effect of temperature on the permeability of beetroot cell membrane

The colorimeter was set up so that it shone green light through the tubes. This is because green is complementary to red. A red solution absorbs most of the green light that is shone on it.

First of all, the student put a tube of distilled water into the colorimeter. This is known as a 'blank'. He set the colorimeter so that this tube gave a reading of 0% absorption. The blank is necessary so that any light absorbed by the tube itself, or the distilled water, is ignored. Therefore any light absorbed by the other tubes will be absorbed by the red pigment.

The student put each tube in turn into the colorimeter and measured the absorbance. The results are shown in Figure 56.

Notice that the permeability of the membrane increases with temperature, but it only increases slowly until the temperature rises above about 50°C when the permeability increases sharply. This is because the proteins in the cell membrane start to denature, allowing pigment to pass out of the cell. Also, as temperature increases, the phospholipids have more kinetic energy and become more fluid. This also allows the membrane to become more 'leaky'.

4  If the student did not have a colorimeter available, how else could he measure the colour of each tube?

Turn to page 91 for the answers.

## Summary

- The cell membrane is composed of a bilayer of phospholipids together with proteins, glycoproteins and glycolipids. This is the fluid-mosaic model of membrane structure.
- Movement across membranes occurs by:
  - simple diffusion of small and/or lipid-soluble molecules across the phospholipid bilayer
  - facilitated diffusion, which involves carrier proteins and channel proteins
  - osmosis, which is the diffusion of water across a membrane down its water potential gradient
  - active transport, which involves carrier proteins and the hydrolysis of ATP
  - co-transport, which involves a carrier protein moving two molecules or ions at the same time

# Cell recognition and the immune system

## Antigens

Each type of cell has large molecules such as proteins and glycoproteins in its cell-surface membrane. These identify the cell and are called **antigens**. They are important in cell recognition. Our immune system can recognise the antigens on its own body cells and identifies them as 'self'. Normally the immune system does not respond to 'self' cells. However, the immune system does recognise antigens that are foreign to the body as 'non-self'. These non-self antigens might be:

- on the surface of pathogens (disease-causing microorganisms) such as bacteria and fungi
- on cells from another organism of the same species, for example, in an organ transplant
- on abnormal body cells, for example, cancer cells
- toxins, for example, those produced by an invading organism, or in a bite from an animal such as a venomous snake

If a pathogen enters the body, its antigens are recognised as 'non-self' by white blood cells called **phagocytes**. These engulf the pathogen by a process called **phagocytosis**. Lysosomes fuse with the vacuole containing the pathogen, releasing their enzymes (lysozymes) and digesting it (Figure 57). The antigens from the pathogen move into the membrane of the macrophage. This is now called an **antigen-presenting cell**.

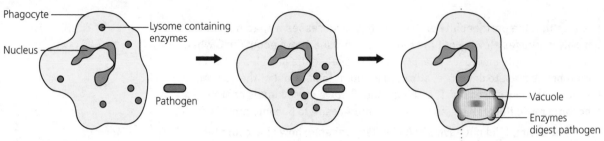

**Figure 57** Phagocytosis

## T lymphocytes: the cellular response

T lymphocytes are another kind of white blood cell. There are many kinds of T lymphocyte in the body, all with receptors on their surface, but each T lymphocyte has a different-shaped receptor.

When an antigen-presenting cell enters the lymph nodes, one of the many different T lymphocytes will have a surface receptor that is complementary to the antigen. The antigen binds to the T lymphocyte's receptor and this **activates** the T lymphocyte. It starts to divide to produce several different **clones** of itself. Figure 58 shows this process.

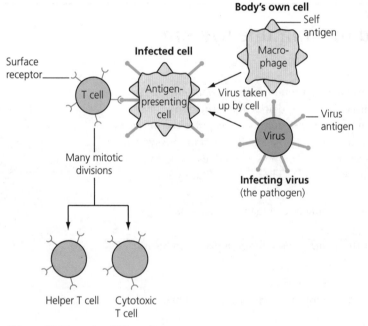

**Figure 58** The role of T lymphocytes

The most important clones produced are:
- helper T cells ($T_H$ cells — see below)
- cytotoxic T cells ($T_C$ cells), which destroy any cells that carry the specific antigen

# B lymphocytes: the humoral response

B lymphocytes are also white blood cells. Like T lymphocytes, they have receptors on their cell-surface membranes. Each B lymphocyte has a different-shaped receptor. The antigen binds to a B lymphocyte with a complementary receptor. The B lymphocyte is then stimulated by a helper T lymphocyte. This is called **clonal selection**. The B lymphocyte divides to form the following clones (Figure 59):

- Plasma cells — these release specific antibodies (see below) into the blood, which bind to the antigen. This leads to the destruction of the antigen or the cell carrying the non-self antigen.
- Memory cells — these carry an immunological memory of the specific antigen. If the same antigen is encountered again on a future occasion, the memory cells will divide and produce specific plasma cells much more quickly than the first time.

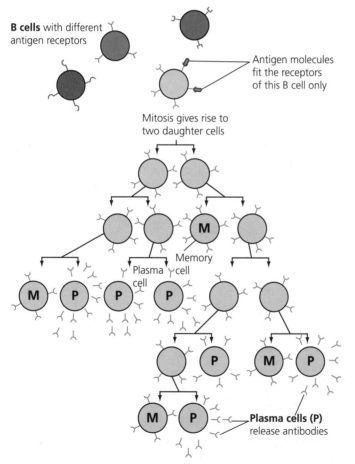

**Figure 59** The cloning of B lymphocytes

An **antibody** (Figure 60) is a protein secreted by a plasma call in response to a non-self antigen.

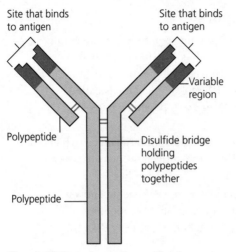

**Figure 60** Structure of an antibody

You will see that the antibody is Y-shaped. All antibodies are the same shape except for the two regions at the end — the variable region. The variable region is complementary to just one antigen, so that the antibody binds to the antigen, forming an antigen–antibody complex. This is the first stage in destroying the non-self antigen or a cell carrying a non-self antigen. The antibodies bind to the antigens, forming a complex of antigens and antibodies. This is called **agglutination**. The complex is then engulfed by phagocytes.

### Primary and secondary immune responses

If you have suffered from a particular infection once you will not catch the disease again. This is because you produced specific memory B lymphocytes as the result of the first infection (Figure 61).

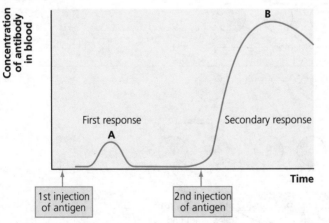

**Figure 61** Primary and secondary immune response

The graph in Figure 61 shows that when you are first exposed to an antigen, there is a short time delay before the antibody concentration in the blood rises. This is because it takes time for the B lymphocytes to become activated, and then form clones of plasma cells that secrete antibodies. However, this **primary response** results in enough antibodies to destroy the infection.

If you encounter the same infection at a later date, a **secondary response** occurs. This time, a huge number of antibodies is secreted almost immediately. This response is much faster because memory B lymphocytes are already present. These divide to form plasma cells that secrete a high concentration of antibodies into the blood. These destroy the pathogen so quickly that the person does not even feel ill.

## Vaccines

It is possible suffer from some kinds of infection more than once, for example, colds. This is because the cold virus keeps changing its antigens, so antibodies against one cold virus will not bind to the antigens of a different strain of the virus.

This secondary response is the basis of **vaccination.** A **vaccine** is a preparation of antigens that is taken into the body to stimulate the production of memory B lymphocytes.

Obviously it would be dangerous to give a person a dangerous pathogen. Therefore, vaccines are made harmless in various ways:

- Giving the person the dead or inactivated pathogen, so it cannot cause disease symptoms, but the antigens are unaltered.
- Using a live but harmless strain of the pathogen. This is an **attenuated** vaccine.
- Separating the antigens from the rest of the pathogen and administering the antigens only.

Vaccines can be used to prevent people ever suffering from serious, life-threatening diseases. It is even possible, if enough people are vaccinated against a disease, to prevent the disease spreading in a population. It is not necessary for everyone in a population to be vaccinated for this to happen. The idea of having enough people vaccinated against a disease so that the disease cannot spread is called **herd immunity**. If enough people in a population are vaccinated, but an unvaccinated person catches the disease, this will mean that the sick person cannot pass the disease on (Figure 62).

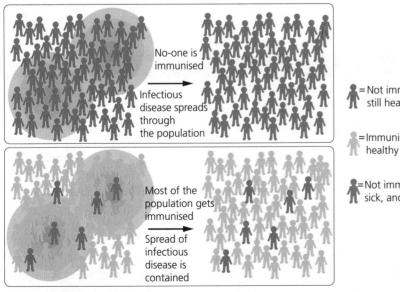

**Figure 62** Herd immunity

**Exam tip**

Do not describe a vaccine as a 'safe dose' of an antigen or, worse still, as a 'safe dose of a disease'. It is a preparation of antigens, but in a safe form to administer.

There are some ethical issues concerning vaccines. While vaccines are generally extremely safe, there is still a low probability of side-effects. Side-effects are most likely to occur in people who have other conditions, for example, epilepsy, so people with conditions that make them more at risk may not receive the vaccine. This is another reason why herd immunity is important.

Another issue is whether everyone should get the vaccine, or only people most at risk. For example, the vaccine to protect against HPV (human papilloma virus) is only given to girls, because HPV is a cause of cervical cancer. Since boys cannot suffer from cervical cancer they are not given the vaccine. However, HPV is sexually transmitted, so some people argue that boys should be vaccinated too, to reduce the chances of boys passing the virus to an unvaccinated girl.

## Active and passive immunity

**Active immunity** is the process just described. In other words, active immunity is when a person is exposed to an antigen and their T and B lymphocytes mount an immune response, resulting in the presence of memory B lymphocytes. However, there are cases when people can receive ready-made antibodies that protect them from disease. This is called **passive immunity**. An example is when an unborn fetus receives antibodies from its mother across the placenta. Another example is when people are injected with ready-made antibodies from animals such as horses or rabbits to treat venomous bites. Passive immunity is different from active immunity because the person is not making the necessary antibodies or memory cells.

> **Knowledge check 38**
>
> Some vaccines are given more than once, for example the MMR vaccine is given in more than one dose. Explain why.

> **Knowledge check 39**
>
> Vaccination cannot be used to treat a person once they have already been infected with a pathogen. Explain why.

> **Knowledge check 40**
>
> A newborn baby receives antibodies from its mother in breast milk. Is this active or passive immunity? Explain why.

> **Knowledge check 41**
>
> A person is vaccinated against TB. Is this active or passive immunity? Explain why.

> **Knowledge check 42**
>
> At one time, antibodies from immunised horses were used to treat people suffering from tetanus. This treatment would not be suitable to use on the same person more than once. Explain why.

## Summary

- Each type of cell has specific molecules on its surface, called antigens, that identify it. These molecules include proteins and enable the immune system to identify:
  - pathogens
  - cells from other organisms of the same species
  - abnormal body cells
  - toxins
- Some pathogens, for example, the cold virus, keep changing their antigens. This means that you can suffer from the same disease again and again, and it also makes it difficult to develop a vaccine.
- T lymphocytes produce a cellular response.
- B lymphocytes produce a humoral response.
- The primary response occurs the first time a person encounters an antigen. This is when plasma cells and memory B cells, specific to the antigen, are produced
- The secondary response occurs the second time a person encounters an antigen. This time antibody production is much faster, quicker and longer-lasting, as a result of memory B cells.
- Vaccines are used to provide protection for individuals and populations against disease. A vaccine is a preparation of antigens in a safe form that can be administered to a person, by injection or otherwise, to stimulate the production of memory B cells.

# Human immunodeficiency virus (HIV)

The structure of the human immunodeficiency virus (HIV) is shown in Figure 63.

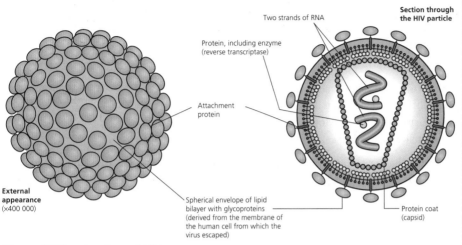

**Figure 63** The structure of HIV

You will see that it has RNA as its genetic material and a membrane around its capsid. It also contains an enzyme called **reverse transcriptase**. HIV is passed from one person to another by transferring body fluids containing the virus, for example by sexual contact or contaminated blood transfusions. The virus specifically infects helper T cells, where it replicates (Figure 64).

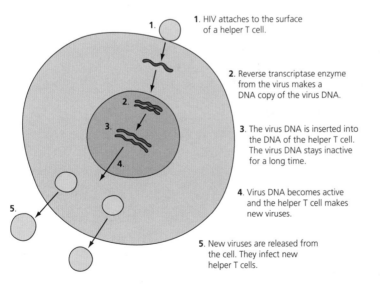

1. HIV attaches to the surface of a helper T cell.

2. Reverse transcriptase enzyme from the virus makes a DNA copy of the virus DNA.

3. The virus DNA is inserted into the DNA of the helper T cell. The virus DNA stays inactive for a long time.

4. Virus DNA becomes active and the helper T cell makes new viruses.

5. New viruses are released from the cell. They infect new helper T cells.

**Figure 64** How HIV infects helper T cells

When HIV first infects helper T cells, the enzyme reverse transcriptase is used to make a DNA copy of the viral RNA. This is inserted into the host cell's DNA. It can remain dormant for a while, so the infected person will have no obvious symptoms, although antibodies will be present shortly after infection. At this stage a person is said to be **HIV positive**. However, after a while, the viral DNA becomes active and the helper T cells start to make new HIV particles. These bud out from the cell, eventually destroying it. These HIV particles go on to infect new helper T cells.

Now that the virus is actively destroying helper T cells, the person's immune system is not working properly. The infected person starts to suffer from diseases that would not normally cause a problem if the immune system was working properly. These infections are called **opportunistic diseases**. Among the diseases that can cause a problem at this stage are Kaposi's sarcoma, which is a form of skin cancer, and TB (tuberculosis).

There are few drugs that can treat viral diseases because they do not show any activity until they are in a living cell. This means that any antiviral drug has to enter the host cell, where it may cause harm to the host. Most antiviral drugs interfere with the process by which the host cell makes new viruses.

Antibiotics are highly effective against bacteria. They kill bacteria by interfering with their metabolism, for example, by inhibiting respiration or protein synthesis, or stopping new cell walls being made. Viruses do not have any metabolism of their own, so antibiotics will not destroy them.

## Monoclonal antibodies

Monoclonal antibodies are antibodies that have come from a single clone of plasma cells. In other words, they are identical antibodies that are complementary to a single, specific antigen. They are very important in medicine. One use of monoclonal antibodies is to target medication to one specific cell type. For example, many cancer drugs are highly toxic and will kill healthy cells as well as cancer cells. If the drug is bound to a monoclonal antibody that binds to a protein receptor on a cancer cell, but not to receptors on healthy cells, this means that the drug can be targeted at cancer cells while doing minimal damage to other cells (Figure 65).

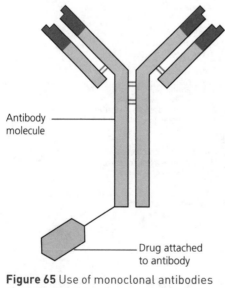

Antibody
molecule

Drug attached
to antibody

**Figure 65** Use of monoclonal antibodies
in cancer treatment

Another use of monoclonal antibodies is in medical diagnosis. Monoclonal antibodies that are specific for particular antigens can be used in immunoassay, to find out the

**Knowledge check 43**

How does destruction of helper T cells lead to an immune system that does not work properly?

**Knowledge check 44**

How does the HIV virus gain entry to helper T cells but not any other kind of cell?

concentration of antibodies a patient has in their blood. They can also be used in diagnostics, for example, to detect blood cancers and to test whether the patient is responding to the therapy they are receiving.

Monoclonal antibodies are also used in ELISA tests. ELISA stands for enzyme-linked immunosorbent assay (Figure 66).

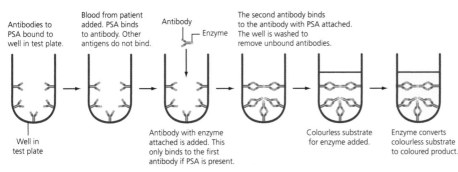

**Figure 66** Using monoclonal antibodies to test for prostate cancer

The example in Figure 66 is testing for prostate-specific antigen (PSA). This test is used to screen for prostate cancer. Similar tests can be used to detect whether a person is HIV positive, or whether a woman is pregnant.

There are some ethical issues involved with monoclonal antibody production. One issue is that it involves the use of mice, and to produce cells that multiply and produce antibodies, cancer cells are needed as well as plasma cells. This means deliberately inducing cancer in mice. Secondly, to make the antibodies suitable for use in humans, transgenic mice are used. A human gene has been inserted into the mice by genetic engineering. Some people have ethical objections to the use of genetic engineering. Thirdly, while monoclonal antibodies have been used successfully to treat many diseases, there have been some problems, both in using them to treat disease and in drug trials.

### Knowledge check 45

Why does a human gene need to be inserted into mice so that they make antibodies suitable for human use?

### Knowledge check 46

In Figure 66:
**a** Why is the well washed at different stages in the test?
**b** Why does the second antibody have an enzyme attached to it?

## Summary

- The human immunodeficiency virus (HIV) contains RNA as its genetic material, within a capsid that is surrounded by a membrane. It replicates inside helper T cells.
- HIV destroys helper T cells, which means that the immune system does not work properly. This leads to opportunistic infections.
- Antibiotics are ineffective against viruses because viruses only replicate inside host cells and have no metabolism of their own.
- Monoclonal antibodies are used in:
  - targeting medication to specific cell types by attaching a therapeutic drug to an antibody
  - medical diagnosis

# Questions & Answers

## Exam format

If you are taking AS Biology, your exams will be structured as follows:

| Paper 1 | Paper 2 |
| --- | --- |
| Any content from topics 1–4, including relevant practical skills | Any content from topics 1–4, including relevant practical skills |
| Written exam, 1 hour 30 minutes<br>75 marks, worth 50% of AS | Written exam, 1 hour 30 minutes<br>75 marks, worth 50% of AS |
| 65 marks: short-answer questions<br>10 marks: comprehension question | 65 marks: short-answer questions<br>10 marks: extended response question |

If you are taking A-level Biology, your exams will be structured as follows:

| Paper 1 | Paper 2 | Paper 3 |
| --- | --- | --- |
| Any content from topics 1–4, including relevant practical skills | Any content from topics 5–8, including relevant practical skills | Any content from topics 1–8, including relevant practical skills |
| Written exam, 2 hours<br>91 marks, worth 35% of A-level | Written exam, 2 hours<br>91 marks, worth 35% of A-level | Written exam, 2 hours<br>78 marks, worth 30% of A-level |
| 76 marks: mixture of long- and short-answer questions<br>15 marks: extended response | 76 marks: mixture of long- and short-answer questions<br>15 marks: comprehension | 38 marks: structured questions, including practical techniques<br>15 marks: critical analysis of experimental data<br>25 marks: essay from a choice of two titles |

## Tips for answering questions

Use the mark allocation. Generally, 1 mark is allocated for one fact, concept or item in an explanation. Make sure your answer reflects the number of marks available.

Respond appropriately to the command words in each question, i.e. the verb the examiner uses. The terms most commonly used are explained below.

- **Describe** — this means 'tell me about…' or, sometimes, 'turn the pattern shown in the diagram/graph/table into words'; you should not give an explanation.
- **Explain** — give biological reasons for *why* or *how* something is happening.
- **Calculate** — add, subtract, multiply, divide (do some kind of sum!) and show how you got your answer — *always* show your working!
- **Compare** — give similarities *and* differences between…
- **Complete** — add to a diagram, graph, flowchart or table.
- **Name** — give the name of a structure/molecule/organism etc.

- **Suggest** — give a plausible biological explanation for something; this term it is often used when testing understanding of concepts in an unfamiliar context.
- **Use** — you must find and include in your answer relevant information from the passage/diagram/graph/table or other form of data.

# About this section

This section contains questions similar in style to those you can expect to see in your exam. The limited number of questions in this guide means that it is impossible to cover all the topics and all the question styles, but they should give you a flavour of what to expect. The responses that are shown are students' answers to the questions.

The papers have the same number of marks as A-level papers 1 and 2.

- Test paper 1 has questions similar to AS paper 2, although it is a little longer than an AS paper. There is no comprehension question, as it is unlikely that a comprehension question would cover only topics 1 and 2.
- Test paper 2 is similar in style to A-level paper 1. No questions typical of paper 3 have been included here as they are synoptic, and you haven't covered enough topics at this stage.

There are several ways of using this section. You could:

- hide the answers to each question and try the question yourself. It needn't be a memory test — use your notes to see if you can actually make all the points you ought to make
- check your answers against the students' responses and make an estimate of the likely standard of your response to each question
- check your answers against the accompanying comments to see where you might have failed to gain marks
- check your answers against the terms used in the question — for example, did you *explain* when you were asked to, or did you merely *describe*?

# Comments

Each question is followed by a brief analysis of what to watch out for when answering the question (icon ⓔ). Student responses are then followed by detailed comments. These are preceded by the icon ⓔ and indicate where credit is due. In the weaker answers, they also point out areas for improvement, specific problems and common errors, such as lack of clarity, weak or non-existent development, irrelevance, misinterpretation of the question and mistaken meanings of terms.

# ■ Test paper 1

This paper has 91 marks available and should take 2 hours to complete. (Your actual AS papers will each have 75 marks available and will take 1 hour 30 minutes to complete.)

## Question 1

Figure 1 shows an *Amoeba*, which is a single-celled organism.

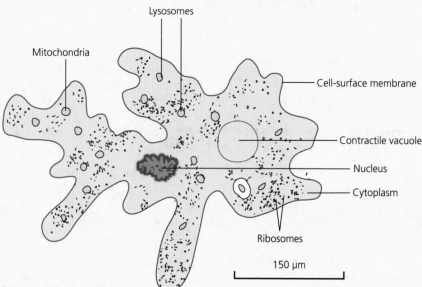

**Figure 1**

**(a) (i)** Calculate the magnification of Figure 1. Show your working. (2 marks)

**(ii)** Give two pieces of evidence from Figure 1 that enable you to identify this as a eukaryotic cell. (2 marks)

**(b)** *Amoeba* lives in freshwater ponds and rivers. It has a contractile vacuole that expels water from the cell. Explain why this is needed. (3 marks)

ⓔ The first one or two questions on a paper are intended to be straightforward. In (a)(i) you are asked to show your working. This is so that you can get credit for the method even if the answer is wrong. Figure 1 has a scale bar, so this should be used to calculate the magnification. In (a)(ii) there is no need to write a lot — just two pieces of evidence are needed, but they must come from Figure 1. Part (b) is asking you to apply your knowledge. The examiner doesn't expect you to have studied *Amoeba* but expects you to work out the answer.

---

**Student A**

**(a) (i)** scale bar measures 30 mm = 30 000 μm

$$\text{magnification} = \frac{\text{measured size}}{\text{real size}}$$

$$= \frac{30\,000}{150} = 200 \checkmark\checkmark$$

**(ii)** The cell has a nucleus ✓ and mitochondria ✓.

ⓔ **4/4 marks awarded** For (a)(i) the right answer is reached, but even if the arithmetic had been wrong, a mark would have been awarded for the correct method. For (a)(ii) these are both correct features of eukaryotic cells, visible in the diagram.

> **(b)** The amoeba has a lower water potential than the freshwater pond ✓ so it takes in water by osmosis ✓. The excess water must be removed from the cell to stop it bursting ✓.

ⓔ **3/3 marks awarded** 1 mark is for stating that the cell has a lower water potential than the pond water, 1 mark for water entering by osmosis, and 1 mark for the reason why water needs to be expelled.

**Student B**

**(a) (i)** scale bar = 3 cm; 3000/150 = 20 ✗✗

**(ii)** It does not have a cell wall ✗ and it has a nucleus ✓.

ⓔ **1/4 marks awarded** For (a)(i) student B has measured in cm, not mm, and then multiplied by 1000. But there are 1000 μm in 1 mm, so this is the wrong conversion. The answer is wrong and so is the working. In (a)(ii) 1 mark is given for mentioning the nucleus. However, some eukaryotic cells (e.g. plant cells) have cell walls, so this is not a feature to distinguish prokaryotic cells from eukaryotic cells.

> **(b)** This stops the cell bursting ✓ as it takes in water by osmosis ✓.

ⓔ **2/3 marks awarded** 1 mark is awarded for saying water enters by osmosis and 1 mark for saying that it stops the cell bursting. However, there is no mention of the water potential gradient, so the third mark cannot be given.

# Question 2

**(a)** Describe a test you could carry out to test for a non-reducing sugar in a solution. (4 marks)

**(b)** Figure 2 shows the structure of sucralose. It is an artificial sweetener that cannot be digested in the body. It is used in low-calorie foods.

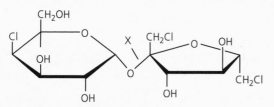

**Figure 2**

**(i)** Name the bond labelled X. (1 mark)

**(ii)** Sucralose contains the same monosaccharides as sucrose. Name these monosaccharides. (1 mark)

**(iii)** Explain why sucralose cannot be digested in the body. (2 marks)

ⓔ The first part of this question is simple recall of knowledge. The second part is asking you to apply your knowledge to an unfamiliar example.

**Student A**

**(a)** First I would do the Benedict's test for reducing sugar and show that this is negative ✓. Then I would boil some of the solution with dilute hydrochloric acid, and then neutralise it with alkali ✓. I would add Benedict's reagent, which is blue, and heat it in a boiling water bath ✓. If a non-reducing sugar is present there will be an orange-red precipitate ✓.

ⓔ **4/4 marks awarded** This is an excellent answer. Student A gives full details and has remembered to do a reducing sugar test first.

**(b) (i)** glycosidic ✓

    **(ii)** glucose and fructose ✓

    **(iii)** It is a different shape from sucrose ✓, so there is no enzyme with an active site that it can fit into ✓.

ⓔ **4/4 marks awarded** Notice that just the word 'glycosidic' is needed to get the mark in (b)(i) — there is no need to write more. (b)(ii) is correct and gets the mark. In (b)(iii) student A mentions shape and fit, and uses the term 'active site'.

**Student B**

**(a)** I would boil some of the solution with acid to hydrolyse it. Then I would add some Benedict's solution and boil it again ✓. If it goes brick red there is reducing sugar present ✓.

ⓔ **2/4 marks awarded** Student B does not remember to do a reducing sugar test first. Also, although an acid hydrolysis has been carried out, there is no mention of neutralising it with alkali before adding the Benedict's reagent.

**(b) (i)** glycosidic ✓

    **(ii)** glucose and galactose

    **(iii)** It is the wrong shape to fit in the enzyme.

ⓔ **1/4 marks awarded** Answer (b)(i) gets the mark. In (b)(ii), glucose is right but galactose is not, so the mark cannot be awarded. (b)(iii) is too vague. There is no reference to the active site of the enzyme and it is not clear that sucralose is a different shape from the molecules that are normally present in the gut.

# Question 3

**(a)** Give two differences between active transport and facilitated diffusion.     (2 marks)

**(b)** Figure 3 shows the concentrations of some ions in pondwater, and inside the cells of a plant that lives in a pond.

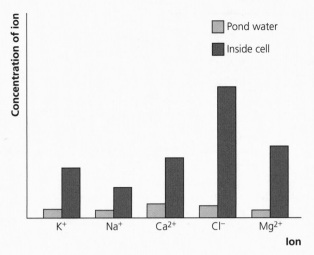

**Figure 3**

(i) How have these ions entered the plant cells? Explain your answer. (2 marks)

(ii) How could you carry out an investigation to check whether your answer is right? (2 marks)

*e* Part (a) is straightforward recall. Part (b) is testing your understanding.

---

**Student A**

**(a)** Active transport uses energy from hydrolysis of ATP but facilitated diffusion is passive and does not require additional energy ✓.

Active transport transports substances against a concentration gradient but facilitated diffusion transports substances down a concentration gradient ✓.

---

*e* **2/2 marks awarded** These answers are excellent and use scientific terminology well.

---

**(b) (i)** Active transport ✓ because all the ions are in a higher concentration inside the cell than outside, so they must have entered against a concentration gradient ✓.

**(ii)** I would grow the plant in the presence of a respiratory inhibitor so it could not produce ATP in respiration ✓. If the ions no longer reached a higher concentration inside the cell than in the pondwater then I would know I was right ✓.

---

*e* **4/4 marks awarded** There is a clear explanation for why these ions have entered the cell by active transport. There is also a clear understanding that inhibiting respiration would reduce the supply of ATP and therefore would reduce active transport.

**Student B**

**(a)** Active transport goes against a concentration gradient but facilitated diffusion goes down a concentration gradient ✓.

Active transport requires energy but facilitated diffusion does not.

Active transport uses carrier proteins but facilitated diffusion does not ✗.

*e* **1/2 marks awarded** The first point here is right and gets a mark. The second point is not well made but would have just gained credit (it should really say that energy from ATP is needed). However, despite the instruction to give two differences, the student has given a third response and it is wrong. Therefore this cancels out the mark that would have been given for the second point here.

**(b) (i)** Using a protein carrier, because ions cannot pass through the phospholipids.

**(ii)** Use a high temperature to denature the proteins and see if the ions can still enter the cell.

*e* **0/4 marks awarded** In (b)(i) it is right to say that ions must enter via a protein, but this is not a method of entry. Student B should have decided whether it was facilitated diffusion or active transport. Also there is no use of information from the graph. Part (b)(ii) is also wrong. Student B has not seen the clue in the first part of the question — that this is to do with either active transport or diffusion. This incorrect answer is clearly a guess.

# Question 4

**(a)** Figure 4 shows a cell undergoing mitosis. Describe what will happen in the next stage.

(3 marks)

Figure 4

**(b)** Figure 5 shows the changes in the mass of DNA during two cell cycles.

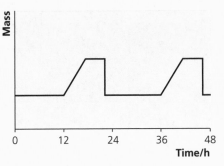

**Figure 5**

   **(i)** On Figure 5, write the letter D to indicate a time when DNA replication is taking place. (1 mark)

   **(ii)** On Figure 5, write the letter M to indicate a time when prophase and metaphase are taking place. (1 mark)

**(c)** Figure 6 shows the structure of cytarabine, a drug that is used to treat cancer. It also shows the structure of deoxycytidine, which forms parts of cytosine-containing nucleotides in DNA. Suggest how cytarabine is effective in treating cancer. (3 marks)

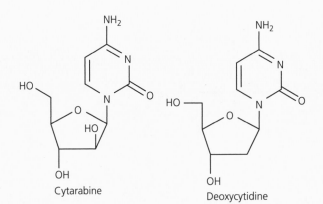

Cytarabine

Deoxycytidine

**Figure 6**

 This question is also testing your knowledge and understanding. Again, part (a) is more straightforward than part (b). Note that when a biology exam question shows you chemical structures, as in part (c), this is usually so that you can see that two molecules are similar in structure. You are not expected to have any detailed knowledge of chemical structures.

**Student A**

**(a)** The centromeres will divide, separating the chromosomes into chromatids ✓. Spindle fibres will contract, pulling ✓ the chromatids towards the poles of the cell ✓.

**e** **3/3 marks awarded** This is an excellent answer. Student A even gives detail about the centromere dividing.

**(b)**

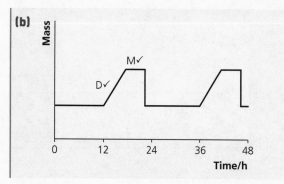

**e** **2/2 marks awarded** Student A gets both marks here.

**(c)** Cytarabine will form part of <u>nucleotides</u> that will be used instead of normal cytosine nucleotides when DNA is replicated ✓. However, it is the wrong shape to fit in the enzyme's active site ✓, so a new molecule of DNA will not be made. This will stop the cell dividing, and cancer cells divide too fast ✓.

**e** **3/3 marks awarded** This is a full answer, gaining full marks. Student A clearly understands that the drug will result in nucleotides that mimic cytosine nucleotides but are slightly different in shape, so DNA polymerase will not be able to build them into functional DNA molecules. The idea that cancer cells divide rapidly, so that stopping DNA replication will stop cancer spreading, is well understood.

### Student B

**(a)** The chromatids will separate ✓ and move to opposite poles of the cell ✓.

**e** **2/3 marks awarded** This answer does not mention the role of the spindle fibres.

**(b)**

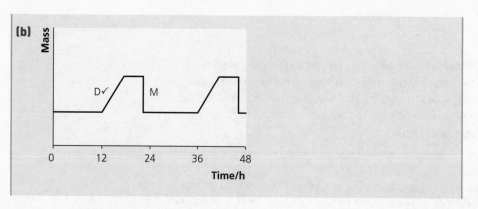

**e** **1/2 marks awarded** Student B has correctly identified when DNA replicates but not mitosis. The downward part of the graph represents cytokinesis when mitosis has finished.

**(c)** These molecules are similar in shape. It will make faulty cytosine nucleotides ✓ that mean the DNA will not work properly.

ⓔ **1/3 marks awarded** This answer scores 1 mark for the idea that the drug will be incorporated into nucleotides that are then slightly different in shape. However there is no mention of fitting into an enzyme's active site, and it is not clear how the DNA will be faulty — this is too vague. Student B does not relate the answer to cancer cells dividing too quickly, either.

# Question 5

**(a)** Complete Table 1 to give the names of the enzymes that catalyse each reaction.    (3 marks)

| Reaction | Enzyme |
|---|---|
| Unwinding DNA and breaking its hydrogen bonds | |
| Hydrolyses lactose into glucose and galactose | |
| Joins DNA nucleotides together in a condensation reaction | |

**Table 1**

A student carried out an investigation to find the optimum pH of the enzyme amylase. She made a plate containing starch agar and cut circular wells in it using a cork borer. She set up the wells so that each contained the same volume and concentration of amylase enzyme in a buffer of different pH values. The plate was left at room temperature for 24 hours.

After this time, the plate was flooded with iodine/potassium iodide solution. Where starch was present, a blue-black colour was seen. The results are shown in Figure 7.

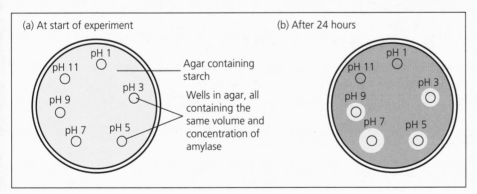

**Figure 7**

**(b)** There were no clear zones round the wells for pH 1 and pH 11. Explain why.    (3 marks)

**(c) (i)** The student concluded that the optimum pH for this enzyme is pH 7. Is this a valid conclusion? Explain your answer.    (2 marks)

**(ii)** Suggest a suitable control for this investigation, and explain why it is needed.    (2 marks)

ⓔ This questions starts with the usual straightforward recall part, but goes on to test how well you can interpret information from an investigation. You will probably have done at least one investigation to test the effect of a variable on the action of an enzyme, but this exact example is likely to be unfamiliar. However, if you understand the principles you should be able to cope with this fairly easily.

**Student A**

**(a)**

| Reaction | Enzyme |
|---|---|
| Unwinding DNA and breaking its hydrogen bonds | DNA helicase ✓ |
| Hydrolyses lactose into glucose and galactose | lactase ✓ |
| Joins DNA nucleotides together in a condensation reaction | polymerase |

ⓔ **2/3 marks awarded** The third enzyme isn't wrong, just vague. Student A should have written 'DNA polymerase'. Note that it is important that the examiner can read that you have written 'lactase' rather than 'lactose'. If your writing is not clear, you will not get the benefit of the doubt. If you have poor handwriting, writing the answer to a question like this in capital letters might be a good idea.

**(b)** The starch had not been hydrolysed ✓ because these pHs are too acid or too alkaline ✓.

The enzyme has been denatured ✓ so it cannot break down the starch.

ⓔ **3/3 marks awarded** All 3 marks have been awarded for clear, correct answers.

**(c) (i)** This is wrong because the student didn't test enough pH values. The enzyme did work better at pH 7 than any other pH tested because it has the largest clear zone ✓. However, the student should now test other pH values between 6 and 8 because one of these might work better ✓.

**(ii)** Repeating this with boiled enzyme and buffer at each pH ✓. This would show it is the enzyme digesting the starch and not the buffer solutions ✓.

ⓔ **4/4 marks awarded** These are excellent answers. Note that there is no mark for saying whether the conclusion is right or not — the marks are for reasons. Student A is clear on why a control is needed and what it should be.

**Student B**

**(a)**

| Reaction | Enzyme |
|---|---|
| Unwinding DNA and breaking its hydrogen bonds | DNA helicase ✓ |
| Hydrolyses lactose into glucose and galactose | disaccharidase |
| Joins DNA nucleotides together in a condensation reaction | DNA polymerase ✓ |

**ⓔ 2/3 marks awarded** 'Disaccharidase' is too vague when you are told that the specific substrate is lactose.

> **(b)** These pHs are too far away from the optimum pH ✓ so the enzyme has been denatured ✓ and starch has not been digested ✓.

**ⓔ 3/3 marks awarded** It would have been better to use the word 'hydrolysed' rather than 'digested'.

> **(c) (i)** Yes, because this had the largest clear zone, so it worked best of all the pHs ✓.
>
> **(ii)** Use distilled water instead of the enzyme ✓ to show that it is the enzyme that is digesting the starch ✓.

**ⓔ 3/4 marks awarded** In (c)(i) it would be wrong to say it is the right conclusion, and no mark is given for this. Student B is aware that this is the pH that worked best of those tested but does not recognise that a fuller range of pH values should be tested before reaching this conclusion. In (c)(ii) student B suggests a different control from that suggested by student A, but it is equally valid and the reason given is also right, so both marks are awarded.

# Question 6

**(a)** Give two properties of water that make it important in biology. (2 marks)

**(b)** Figure 8 shows the volume of red blood cells when placed in different concentrations of sodium chloride.

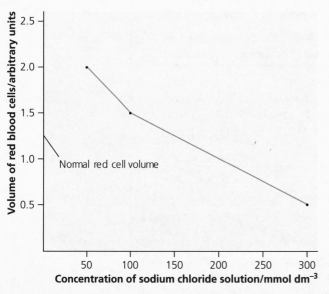

**Figure 8**

(i)   Describe the relationship shown by Figure 8.                                             (1 mark)

(ii)  Use your knowledge of water potential to explain this relationship.     (3 marks)

(iii) Suggest why Figure 8 does not show any values for concentrations of
      sodium chloride below $50\,mmol\,dm^{-3}$.                                          (2 marks)

 The first part of the question is straightforward recall. The rest of the question is harder as you need to apply your knowledge and understanding to what is probably a new example for you. When asked to describe a graph, make sure you do just that — graphs at A-level are rarely straight lines, so you need to describe the trend and also refer to any points where there is a change in gradient. The term 'suggest' is a clue that the answer will not be in your lesson notes, but nevertheless you should be able to apply your A-level knowledge to answering the question.

### Student A

(a) Water has a relatively high heat capacity, making it slow to heat up and cool down ✓.

It is an important solvent in which metabolic reactions take place in cells ✓.

@ **2/2 marks awarded** Two well-expressed points are made.

(b) (i)   As the sodium chloride concentration increases, the volume of the red blood cells decreases. It decreases rapidly at first and then slightly less rapidly from $100\,mmol\,dm^{-3}$ onwards ✓.

(ii)  If the water potential outside the cell is greater than the water potential inside the cell, the cell takes in water by osmosis and swells. When the water potential outside the cell is lower than inside the cell ✓, the cell loses water by osmosis ✓ and it shrinks in volume ✓.

(iii) These concentrations have a very high water potential compared with the cell ✓, so the cell takes in so much water by osmosis that it bursts ✓.

@ **6/6 marks awarded** In (b)(i) the graph is described well and student A has noticed the change in gradient. Part (b)(ii) is a good answer, using the correct terminology. A clear explanation, using water potential terminology, is provided for (b)(iii).

### Student B

(a) It takes part in hydrolysis reactions ✓.

It is good for cooling you down.

@ **1/2 marks awarded** A mark is given for the first point but the second point does not give the property of water that makes it good for cooling you down. Student B should have mentioned the high latent heat of vaporisation.

**(b) (i)** As the concentration of sodium chloride increases, the volume of the cells decreases.

**(ii)** Water leaves the cells by osmosis ✓ when the water potential outside the cell is lower than the water potential inside the cell, and the other way round ✓.

**(iii)** The cells get even smaller and they are too hard to measure ✗.

ⓔ **2/6 marks awarded** In (b)(i) student B just states the general trend and has not noticed the change in gradient, so no marks are awarded. The answer to (b)(ii) gets 2 marks out of 3 — losing a mark for not explaining how this changes the volume of the cells. In (b)(iii) student B has not read the question properly and is looking at the right-hand side of the graph, consequently inventing a rather unlikely explanation. He/she needed to realise that a concentration below $50\,\text{mmol dm}^{-3}$ would be on the left-hand side of the graph.

# Question 7

Figure 9 shows the human immunodeficiency virus.

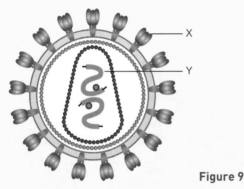

**Figure 9**

**(a)** Name molecule Y. (1 mark)

**(b)** Molecule X is important in causing B cells to produce antibodies against the virus. Explain how. (4 marks)

**(c)** Name two organelles that would become more active in a B cell when it starts producing antibodies. (2 marks)

**(d)** Molecule X can vary in shape. This means that it is difficult to develop a vaccine against HIV. Explain why. (2 marks)

ⓔ This is a straightforward question, testing your knowledge and understanding. In (a) and (c) you can give brief answers to get the marks — do not waste valuable time writing more than you need to.

> ### Student A
> **(a)** RNA ✓

ⓔ **1/1 mark awarded** This is the correct answer.

**(b)** Molecule X acts as an antigen ✓. A B cell in the lymph nodes with the right shape receptor to fit the antigen is activated by a helper T cell ✓. The activated B cell divides to produce a clone of plasma cells ✓ and memory B cells. The plasma cells secrete antibodies that are specific to the viral antigen ✓.

ℯ **4/4 marks awarded** This is a good answer, gaining full marks.

**(c)** ribosomes ✓ and rough endoplasmic reticulum ✓

ℯ **2/2 marks awarded** These are two good answers.

**(d)** If it changes shape, then antibodies against the previous antigen will not be effective against the new shape ✓. This means that the memory cells produced from a vaccine will not produce the correct-shaped antibodies ✓ to fit the new antigen.

ℯ **2/2 marks awarded** This is a good answer, gaining both marks. Student A has referred to the idea of shape and the need for antibodies to bind to the antigen, and also refers to memory cells being produced as a result of vaccination.

## Student B

**(a)** genetic material ✓

ℯ **1/1 mark awarded** This is correct even though the specification does specify more detail than this. However, if you can state that it is RNA, this is a better answer.

**(b)** It triggers a specific B cell to divide and produce plasma cells.✓. These plasma cells secrete antibodies with the correct binding sites to attach to the virus ✓ and destroy it.

ℯ **2/4 marks awarded** The answer does not state that molecule X acts as an antigen and does not refer to helper T cells activating B cells. However, the production of plasma cells and the secretion of specific antibodies are both mentioned.

**(c)** ribosomes ✓ and endoplasmic reticulum

ℯ **1/2 marks awarded** The answer should specify *rough* endoplasmic reticulum.

**(d)** The vaccine will produce antibodies against a specific shape of molecule. If the molecule changes shape, the antibodies will not fit ✓.

ℯ **1/2 marks awarded** Student B has the right idea that changing the shape of molecule X means that previous antibodies will not fit to it. However, there is no reference to memory cells.

# Question 8

Figure 10 shows a drawing of a liver cell as seen using an electron microscope.

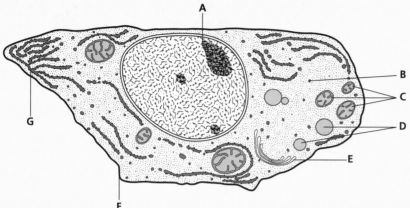

**Figure 10**

**(a)** Complete Table 2 to give the name and function of each structure A–G.    (7 marks)

| Label | Name of structure | Function |
|-------|-------------------|----------|
| A | | |
| B | | |
| C | | |
| D | | |
| E | | |
| F | | |
| G | | |

**Table 2**

**(b)** Give one advantage and one disadvantage of using an electron microscope, rather than an optical microscope.    (2 marks)

**(c)** Give two ways in which prokaryotic cells are different from eukaryotic cells.    (2 marks)

ⓔ Again this is a straightforward question, but note that in (a) you need the name of the organelle *and* its function to gain a mark.

**Student A**

**(a)**

| Label | Name of structure | Function |
|-------|-------------------|----------|
| A | Nucleolus | Synthesises ribosomes ✓ |
| B | Ribosome | Makes proteins ✓ |
| C | Mitochondrion | Aerobic respiration ✓ |
| D | Lysosome | Contains hydrolytic enzymes that destroy worn-out organelles ✓ |
| E | Golgi apparatus | Modifies proteins and packages them for secretion ✓ |
| F | Cell-surface membrane | Controls what enters and leaves the cell ✓ |
| G | Rough endoplasmic reticulum | Synthesises and transports proteins ✓ |

ⓔ **7/7 marks awarded** This is a full set of correct answers.

**(b)** Advantage: the EM has better resolution than an optical microscope ✓.

Disadvantage: the EM is harder to use.

ⓔ **1/2 marks awarded** A clear advantage is given but the disadvantage is too vague. Saying it is harder to use (or more expensive) without explaining clearly will not get you a mark. A mark would have been given for saying that specimens need lengthy preparation for EM, while an optical microscope requires only simple preparation.

**(c)** Prokaryotic cells have cell walls made of murein but if eukaryotic cells have a wall it is usually made of cellulose ✓.

Prokaryotic cells have circular DNA but eukaryotic cells have linear DNA ✓.

ⓔ **2/2 marks awarded** These are two well-expressed answers. Notice that the student has made it clear which cell is being described and a comparison between the two cell types has been made.

# Question 9

**(a) (i)** Why is the cell-surface membrane described as *fluid mosaic*? (2 marks)

**(ii)** What is the function of cholesterol in cell membranes? (1 mark)

**(b)** Give two functions of the proteins in cell surface membranes. (2 marks)

**(c)** In 1925 Gorter and Grendel investigated the structure of cell membranes.

- They dissolved the lipids from all the red blood cells in $1\,mm^3$ of blood in an organic solvent and then spread these lipid molecules on the surface of water where they would form a layer one molecule thick.
- The area occupied by these lipids was $0.92\,m^2$.
- The surface area of all the red blood cells in $1\,mm^3$ of blood is $0.47\,m^2$.

**(i)** What did this tell Gorter and Grendel about the structure of cell membranes? Use suitable calculations to support your answer. (2 marks)

**(ii)** Red blood cells have no internal organelles. Explain the importance of this to the validity of their data. (2 marks)

ⓔ This is mostly testing recall of knowledge. However, the data are intended to test your understanding using an unfamiliar context. Notice that in (c)(i) you are asked to use calculations to support your answer.

---

**Student A**

**(a) (i)** It is fluid because the phospholipid molecules move around and change places with each other ✓ and it is mosaic because the proteins form a mosaic pattern, scattered among the phospholipids ✓.

**(ii)** It makes the membrane more stable by restricting the movement of molecules in the membrane ✓.

ⓔ **3/3 marks awarded** This answer is expressed well.

**(b)** They can act as carrier proteins ✓ or enzymes ✓.

🅔 **2/2 marks awarded** Both answers are correct.

**(c) (i)** It showed them that the cell surface membrane is two lipid molecules thick ✓. This is because the area occupied by the single layer of lipids (0.92 m²) is almost exactly double the total surface area of the red blood cells (0.47 m²) ✓.

**(ii)** If there were internal organelles, some of these would have had membranes around them, such as mitochondria ✓. This means that when they dissolved the lipids from the cell they would have had more molecules than were in the cell-surface membrane ✓.

🅔 **4/4 marks awarded** In (c)(i) 1 mark is for the correct deduction and 1 mark is for the calculation, although saying one figure is almost exactly double another is about the simplest calculation that would be allowed. (c)(ii) is another clear and detailed answer, gaining both marks.

**Student B**

**(a) (i)** It can move around and it is mosaic because the proteins form a mosaic.

**(ii)** It makes the membrane more fluid.

🅔 **0/3 marks awarded** In (a)(i) the student says 'it' moves around, which presumably means the membrane. The membrane does not move significantly — it is the molecules in it that move, so this is wrong. The proteins form a mosaic but as this is the term the student is asked to explain, the answer cannot be credited without explaining that the mosaic is a pattern, or that the proteins are scattered among the phospholipids. In (a)(ii) the answer is wrong because it makes the membrane more stable and restricts movement.

**(b)** cell recognition ✓, carriers ✓ and channels

🅔 **2/2 marks awarded** Student B has given three answers when only two were asked for. Luckily, all three are right, so no marks are lost.

**(c) (i)** Cell membranes are two phospholipids thick ✓. This is because 0.92/0.47 = 1.96, which is nearly 2 ✓.

**(ii)** Some cell organelles are surrounded by membranes ✓.

🅔 **3/4 marks awarded** In (c)(i) both marks are awarded and there is a clear calculation to support the answer. In (c)(ii) there is 1 mark for understanding that some cell organelles have membranes around them, but this is not linked to the validity of the data, so the second mark cannot be awarded.

# Question 10

(a) Describe how HIV causes the symptoms of AIDS. (6 marks)

(b) Describe how T cells respond to a viral infection. (5 marks)

(c) Explain the differences between active and passive immunity. (4 marks)

ⓔ This kind of question tests your knowledge and requires extended prose.

### Student A

(a) HIV infects helper T cells ✓. The virus's genetic information becomes incorporated into the host cell DNA and may stay dormant for a long time. After a period of time, the virus DNA becomes active and the virus multiplies inside these cells ✓, destroying the cells when they burst out. T cells are important in activating B cells ✓, so when T cells are destroyed by HIV the immune system does not work properly ✓. This means the person is susceptible to opportunistic diseases ✓, which are infectious diseases they would probably fight off if they had the normal number of helper T cells ✓. An example is TB.

ⓔ **6/6 marks awarded** This is a full and correctly worded explanation.

(b) The antigen is presented to a T cell in the lymph nodes ✓. The T cell with receptors on its surface that are complementary to the antigen ✓ is activated ✓. This divides to form clones of T cells. One clone is cytotoxic T cells, which destroy any cell that carries the specific antigen ✓. Another clone is helper T cells, which activate B cells and increase antibody production by B cells ✓.

ⓔ **5/5 marks awarded** This answer gives full details in a logical order and uses good scientific terminology.

(c) Active immunity is when the body produces its own memory cells ✓ and plasma cells that release antibodies ✓. Passive immunity is when the person is given ready-made antibodies ✓, for example a baby receiving breast milk. Passive immunity is short lived but active immunity is long lasting.

ⓔ **3/4 marks awarded** This answer just missed the last mark because it does not explain why passive immunity is short term. If student A had added that the antibodies are broken down soon after they are given, the fourth mark could have been awarded.

### Student B

(a) HIV infects T cells. The virus multiplies inside these cells. This reduces the number of T cells ✓. Without enough T cells the person's immune system is not very effective ✓ at stopping the person getting infectious diseases. So the person gets a lot of infectious diseases and if AIDS is not treated properly they will eventually die from an infection ✓.

ⓔ **3/6 marks awarded** Student B says that the virus infects T cells, but does not mention that it infects *helper* T cells and that helper T cells are important in activating B cells. Although the answer does mention that the person is susceptible to infectious diseases, another mark could have been awarded if the term 'opportunistic diseases' had been used.

> **(b)** The antigen activates a specific T cell ✓, which divides to form clones of cytotoxic T cells, helper T cells and plasma cells ✗. The cytotoxic T cells destroy the antigen ✓.

ⓔ **2/5 marks awarded** The answer states that a T cell is activated but does not say that it has specific receptors in its membrane, nor that this is in the lymph nodes. Student B correctly says that cytotoxic T cells are produced, and what they do, but does not say what the helper T cells do. In addition, there is an error — T cells do not produce plasma cells. Plasma cells are produced by B cells.

> **(c)** Active immunity is when a person makes their own antibodies from plasma cells ✓ and makes memory cells ✓. Passive immunity is when a person is injected with antibodies that have been made by an experimental animal. ✓ This is much faster than active immunity because in active immunity it takes time for the plasma cells to produce the antibodies ✓.

ⓔ **4/4 marks awarded** This is a well-expressed answer with plenty of detail. If a fifth mark had been available this would have scored another mark for saying that active immunity is fast-acting.

# ■ Test paper 2

This test has 91 marks available, and should take 2 hours to complete. This is similar in style to A-level paper 1.

## Question 1

**Figure 1 shows some biological molecules.**

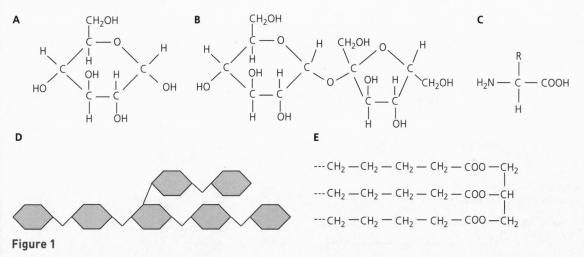

**Figure 1**

# Questions & Answers

**(a)** Give the letter of the molecule(s) that fit each description in Table 1.

(5 marks)

| Description | Molecule(s) |
|---|---|
| A molecule that is a polymer | |
| A molecule that contains a glycosidic bond | |
| A molecule that contains an ester bond | |
| A molecule that would give a positive Benedict's test for reducing sugar | |
| A molecule that joins together with others to form a polypeptide | |

Table 1

**(b) (i)** What is the secondary structure of a protein?

(1 mark)

**(ii)** Name two bonds that hold a protein in its tertiary structure.

(2 marks)

ⓔ This paper, like most, starts with questions that should be fairly easy. However, be careful in part (a) as there might be more than one answer in each case.

## Student A

**(a)**

| Description | Molecule(s) |
|---|---|
| A molecule that is a polymer | D ✓ |
| A molecule that contains a glycosidic bond | B, D ✓ |
| A molecule that contains an ester bond | E ✓ |
| A molecule that would give a positive Benedict's test for reducing sugar | A ✓ |
| A molecule that joins together with others to form a polypeptide | C ✓ |

ⓔ **5/5 marks awarded** Five correct answers. Note that two examples are needed in one case.

**(b) (i)** This is the sequence of amino acids in the polypeptide ✓.

**(ii)** hydrogen bonds ✓ and disulfide bridges ✓

ⓔ **3/3 marks awarded** A succinct answer, for full marks.

## Student B

**(a)**

| Description | Molecule(s) |
|---|---|
| A molecule that is a polymer | D, E |
| A molecule that contains a glycosidic bond | B |
| A molecule that contains an ester bond | E ✓ |
| A molecule that would give a positive Benedict's test for reducing sugar | A, B |
| A molecule that joins together with others to form a polypeptide | C ✓ |

ⓔ **2/5 marks awarded** Unfortunately student B thinks that E, the triglyceride, is a polymer but it is not. So even though the correct answer D is there, there can be no mark given. B contains a glycosidic bond, but so does D, so again there is no mark. Student B does realise that the triglyceride contains an ester bond. Although glucose in A does give a positive reducing sugar test, B does not as it is sucrose, which is a non-reducing sugar. Student B is right in saying that the amino acid, C, is the molecule that forms part of a polypeptide.

**(b) (i)** The order of amino acids it contains ✓.

**(ii)** ionic bonds ✓ and peptide bonds ✗

ⓔ **2/3 marks awarded** Although a protein contains peptide bonds, these hold the protein in its primary structure, not the tertiary structure.

## Question 2

Figure 2 represents the cell cycle.

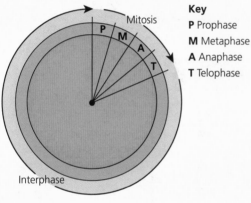

**Figure 2**

**(a) (i)** Put a cross on Figure 2 to show where cytokinesis takes place. (1 mark)

**(ii)** Give two activities that occur in the cell during interphase. (1 mark)

**(b)** Vincristine is a drug that stops spindle fibres forming. What effect would this drug have on mitosis? Explain your answer. (2 marks)

**(c)** Table 2 shows the percentage of nucleotides containing each of the four bases in the DNA from various organisms.

| Source of DNA | Adenine % | Guanine % | Thymine % | Cytosine % |
|---|---|---|---|---|
| Human | 30 | 20 | 30 | 20 |
| Turtle | 28 | 22 | 28 | 22 |
| *E. coli* | 24 | | | |
| Salmon | 29 | 21 | 29 | 21 |
| Virus | 25 | 24 | 33 | 18 |

**Table 2**

(i)   Complete Table 2 to show the figures for *E. coli*. Explain how you arrived at this answer.

(1 mark)

(ii)  The structure of the DNA in the virus is not the same as the structure of DNA in the other organisms. Suggest what this difference in DNA structure might be.

(1 mark)

*e*  This question is testing your knowledge and understanding. It starts with simple recall but then moves to giving you information that is probably new to you, to check whether you can apply your knowledge to an unfamiliar situation.

### Student A

(a) (i)   (cross placed right next to the line after telophase) ✓

(ii)  DNA replication and protein synthesis ✓

*e*  **2/2 marks awarded** Student A clearly understands this topic well.

(b)  Mitosis will not be able to go past prophase ✓, because spindle fibres are needed for the chromosomes to line up on during metaphase ✓.

*e*  **2/2 marks awarded** Student A says clearly what will happen and then gives a correct reason to explain.

(c) (i)

| Source of DNA | Adenine % | Guanine % | Thymine % | Cytosine % |
|---|---|---|---|---|
| Human | 30 | 20 | 30 | 20 |
| Turtle | 28 | 22 | 28 | 22 |
| *E. coli* | 24 | **26** | **24** | **26** |
| Salmon | 29 | 21 | 29 | 21 |
| Virus | 25 | 24 | 33 | 18 |

This is because adenine pairs with thymine, and cytosine with guanine so the amount of adenine nucleotides will be the same as thymine. The rest of the nucleotides will be half cytosine and half guanine ✓.

(ii)  The virus DNA must be single-stranded because A does not equal T and C does not equal G ✓.

*e*  **2/2 marks awarded** This is a precise answer, for full marks.

### Student B

(a) (i)   (a cross in telophase) ✗      (ii)  DNA replication and respiration ✓

*e*  **1/2 marks awarded** No mark is awarded for (a)(i) because cytokinesis happens immediately *after* telophase. However, student B gets a mark for two correct processes in (a)(ii).

(b)  Metaphase will not happen ✓ because a spindle is needed for the chromosomes to attach to ✓.

*e*  **2/2 marks awarded** This is all that is needed for 2 marks.

**(c) (i)**

| Source of DNA | Adenine % | Guanine % | Thymine % | Cytosine % |
|---|---|---|---|---|
| Human | 30 | 20 | 30 | 20 |
| Turtle | 28 | 22 | 28 | 22 |
| *E. coli* | 24 | **26** | **24** | **26** |
| Salmon | 29 | 21 | 29 | 21 |
| Virus | 25 | 24 | 33 | 18 |

**(ii)** The DNA is not double stranded because there is no complementary base-pairing ✓.

*e* **1/2 marks awarded** No mark is awarded for (c)(i) as there is no explanation to support the (correct) numbers in the table. The answer to (c)(ii) is correct.

# Question 3

**(a) (i)** What is a capsid? (1 mark)

**(ii)** Viruses are considered to be non-living. Explain why. (2 marks)

**(b)** HIV infects helper T cells but not any other kind of cell. Explain how. (2 marks)

*e* This is another fairly straightforward question and answers can be brief if they are focused on the right points.

### Student A

**(a) (i)** A protein layer around the genetic material of a virus ✓.

**(ii)** They are thought to be non-living because they do not have a cell structure ✓ and they show no metabolic activity such as respiration ✓.

*e* **3/3 marks awarded** Two correct answers for full marks.

**(b)** HIV has an attachment protein on its outside surface that fits with a receptor protein ✓ in the cell surface membrane of a helper T cell. Other kinds of cell do not have this attachment protein in their cell membranes ✓.

*e* **2/2 marks awarded** This is a fully correct answer that uses the right terminology.

### Student B

**(a) (i)** A protein coat around the virus ✓.

**(ii)** Because they can only reproduce inside living cells ✓ and have no metabolism of their own ✓.

*e* **3/3 marks awarded** Two brief answers, but enough for full marks.

**(b)** HIV fits with a receptor protein that is found only in the membrane of a helper T cell ✓.

*e* **1/2 marks awarded** Student B fails to refer to the attachment protein on the surface of the virus.

# Question 4

Figure 3 shows part of a molecule of DNA.

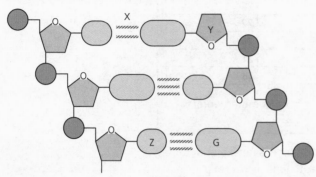

**Figure 3**

**(a)** Name: **(i)** bonds X          **(ii)** molecule Y          **(iii)** molecule Z.          (3 marks)

Meselson and Stahl carried out an investigation to find out whether DNA replication is conservative or semi-conservative. They grew bacteria for many generations in a medium containing a heavy isotope of nitrogen ($^{15}N$). These were then transferred to a medium containing the normal isotope of nitrogen ($^{14}N$).

DNA samples were extracted from the bacteria after they had been growing on the heavy nitrogen medium for many generations (generation 0), after one generation on the normal nitrogen medium (generation 1) and again after two generations on the normal nitrogen medium (generation 2). The DNA was centrifuged in a solution. Figure 4 shows some of the results.

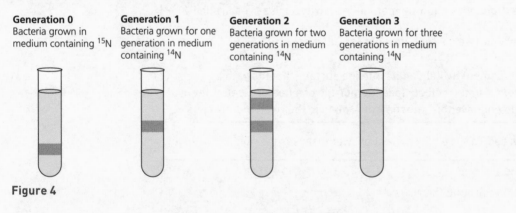

**Generation 0**
Bacteria grown in medium containing $^{15}N$

**Generation 1**
Bacteria grown for one generation in medium containing $^{14}N$

**Generation 2**
Bacteria grown for two generations in medium containing $^{14}N$

**Generation 3**
Bacteria grown for three generations in medium containing $^{14}N$

**Figure 4**

**(b)** **(i)** These results support the hypothesis that DNA replicates semi-conservatively. Explain how.          (3 marks)

   **(ii)** Complete Figure 4 to show the results you would expect after one more generation on the $^{14}N$ medium.          (2 marks)

ⓔ There is a lot of information in the second part of this question about an investigation that is probably unfamiliar to you. The examiner does not expect you to know this investigation, but expects you to apply your knowledge in an unfamiliar situation. Make sure you read the information in the question carefully before writing your answer.

---

**Student A**

**(a) (i)** hydrogen ✓ **(ii)** deoxyribose ✓ **(iii)** cytosine ✓

---

ⓔ **3/3 marks awarded** Precise answers are needed here. Compare student B's response.

**(b) (i)** The DNA in generation 1 is intermediate ✓ in density between generation 0, which is 'heavy', and generation 2, which is 'light'. This means that it must have one strand that is heavy ✓ and the newly formed strand must be light ✓.

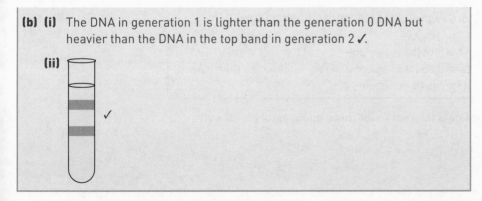

**(ii)**

ⓔ **5/5 marks awarded** This is a fully correct and detailed answer.

---

**Student B**

**(a) (i)** hydrogen ✓ **(ii)** five-carbon sugar **(iii)** organic base

---

ⓔ **1/3 marks awarded** The answers to (a)(ii) and (iii) are not wrong but they are not detailed enough. The question says this is DNA, so the five-carbon sugar must be deoxyribose. Also, the base must be cytosine as it pairs with guanine.

**(b) (i)** The DNA in generation 1 is lighter than the generation 0 DNA but heavier than the DNA in the top band in generation 2 ✓.

**(ii)**

ⓔ **2/5 marks awarded** In (b)(i) student B understands that the DNA is intermediate in density but does not get the mark for explaining that this means it has one 'heavy' strand and one new 'light' strand. In (b)(ii) the bands are in the right places but as there will be more newly formed DNA that is 'light', the upper band will be thicker and the lower band will be thinner.

## Question 5

**A student was given tubes containing the following:**

| Contents of tube | Result of Benedict's test | Result of biuret test |
|---|---|---|
| Amylase and starch that had been incubated for 12 hours | | |
| Albumen, a protein | | |
| Glucose | | |
| Sucrose | | |

Table 3

**(a)** She carried out a Benedict's test for reducing sugars on all the tubes, followed by a biuret test. Complete Table 3 to show the results you would expect for each tube. Use '✓' for a positive result and '✗' for a negative result. (4 marks)

**(b)** Describe the features of glycogen that make it a good storage molecule. (3 marks)

ⓔ This should be straightforward if you have learned your tests for biological molecules.

**Student A**

**(a)**

| Contents of tube | Result of Benedict's test | Result of biuret test | |
|---|---|---|---|
| Amylase and starch that had been incubated for 12 hours | ✓ | ✓ | ✓ |
| Albumen, a protein | ✗ | ✓ | ✓ |
| Glucose | ✓ | ✗ | ✓ |
| Sucrose | ✗ | ✗ | ✓ |

ⓔ **4/4 marks awarded** 1 mark is given for each for each correct row.

**(b)** It is insoluble, so doesn't affect the water potential of the cell ✓. It is compact, so it doesn't take up a lot of space ✓. It is branched, so there are lots of 'ends' to break off glucose molecules ✓.

ⓔ **3/3 marks awarded** This gets full marks for three good features, all well explained.

**Student B**

**(a)**

| Contents of tube | Result of Benedict's test | Result of biuret test | |
|---|---|---|---|
| Amylase and starch that had been incubated for 12 hours | ✓ | ✗ | ✗ |
| Albumen, a protein | ✗ | ✓ | ✓ |
| Glucose | ✓ | ✗ | ✓ |
| Sucrose | ✓ | ✗ | ✗ |

ⓔ **2/4 marks awarded** Student B has not realised that the enzyme amylase is a protein and will therefore give a positive biuret test result, or that sucrose is a non-reducing sugar so will not give a positive result for Benedict's test.

**(b)** It is insoluble and coils up so a lot of glucose is stored in a small space ✓.

ⓔ **1/3 marks awarded** This gets 1 mark for the 'compact' idea, but the statement that glycogen is insoluble does not get a mark without explaining why this is a useful feature.

# Question 6

A student carried out an investigation to find the water potential of onion tissue. She was given $500\,cm^3$ of 1.0 M sucrose solution. She used this to set up two sets of test tubes with sucrose concentrations ranging from 1.0 M to 0.1 M. Each set of tubes contained $20\,cm^3$ of sucrose solution.

**(a)** Complete Table 4 to show the volumes of distilled water and 1.0 M sucrose solution she placed in each tube to produce $20\,cm^3$ of the concentration required.

(2 marks)

| Concentration of sucrose/M | Volume of 1.0 M sucrose added/$cm^3$ | Volume of distilled water added/$cm^3$ |
|---|---|---|
| 0.9 | | |
| 0.4 | | |

Table 4

In one set of tubes she put a thin slice of onion tissue. In the other set of tubes she placed a drop of methylene blue dye. She put a bung in each tube and left them in a cool place for 24 hours.

After 24 hours, the student carefully removed the onion tissue from the tubes containing onion. She replaced the bung and gently tapped the tubes to make sure the contents were mixed.

The student took a Pasteur pipette and removed a few drops of the 1.0 M solution with the blue dye in it. She carefully inserted the Pasteur pipette into the clear solution and released a drop of dye. This is shown in Figure 5.

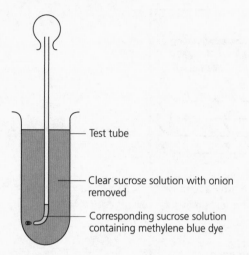

Test tube

Clear sucrose solution with onion removed

Corresponding sucrose solution containing methylene blue dye

**Figure 5**

The student observed carefully whether the blue drop rose, fell or hovered. Her results are shown in Table 5.

| Sucrose solution/M | Movement of blue drop |
|---|---|
| 1.0 | Fell |
| 0.9 | Fell |
| 0.8 | Fell slowly |
| 0.7 | Hovered |
| 0.6 | Rose slowly |
| 0.5 | Rose |
| 0.4 | Rose |
| 0.3 | Rose |
| 0.2 | Rose |
| 0.1 | Rose |

**Table 5**

**(b)** Use your knowledge of water potential to explain the results for:

   **(i)** 1.0 M                                               (3 marks)

   **(ii)** 0.1 M                                           (3 marks)

**(c)** The student decided that the water potential of the onion tissue was similar to the water potential of 0.7 M sucrose. Explain why.    (2 marks)

**(d)** **(i)** Explain why it was important that the student put a bung in each tube before leaving it for 24 hours.    (1 mark)

   **(ii)** It was not important to weigh the slices of onion that the student used. Explain why.    (2 marks)

(e) This question is testing whether you can apply your understanding to what is probably an unfamiliar practical situation. However, if you understand the topic and build on practical skills you have acquired you should be able to answer this. Once again, make sure you read the information in the question carefully.

**Student A**

(a)

| Concentration of sucrose/M | Volume of 1.0 M sucrose added/cm$^3$ | Volume of distilled water added/cm$^3$ |
|---|---|---|
| 0.9 | 18 | 2 ✓ |
| 0.4 | 8 | 12 ✓ |

(e) **2/2 marks awarded** There is 1 mark for each correct row.

(b) (i) The 1.0 M sucrose solution has a lower water potential than the onion cells ✓. Therefore water leaves the onion cells by osmosis ✓. This lowers the density of the solution, so when the drop of dye enters it, it sinks ✓.

(ii) The 0.1 M sucrose solution has a higher water potential than the onion cells ✓. Therefore water enters the onion cells by osmosis ✓. This increases the density of the solution, so when the drop of dye enters it, it rises ✓.

(e) **6/6 marks awarded** These answers show clear understanding and are well expressed.

(c) This is because the solutions are similar in density, as the drop of dye does not rise or fall ✓. Therefore the onion cells and the solution have not taken in or lost any water because they have the same water potential ✓.

(e) **2/2 marks awarded** Full marks for a correct reason here, expressed in terms of water potential.

(d) (i) So no evaporation could take place that would change the water potential ✓.

(ii) The investigation is measuring whether the solution has taken in or lost water ✓. This does not depend on the mass of the tissue used ✓.

(e) **3/3 marks awarded** This is clearly expressed using the term 'water potential', and also correctly refers to evaporation.

**Student B**

(a)

| Concentration of sucrose/M | Volume of 1.0 M sucrose added/cm$^3$ | Volume of distilled water added/cm$^3$ |
|---|---|---|
| 0.9 | 18 | 2 ✓ |
| 0.4 | 4 | 6 |

(e) **1/2 marks awarded** Although the proportions are right in the second row, this would not give a total volume of 20 cm$^3$.

**(b) (i)** Water has left the onion cells by osmosis ✓. This lowers the density of the solution, so when the drop of dye enters it, it sinks ✓.

**(ii)** Water has entered the onion cells by osmosis ✓. This increases the density of the solution, so when the drop of dye enters it, it rises ✓.

ⓔ **4/6 marks awarded** Student B has not used the term 'water potential' in the explanation.

**(c)** This is because the solutions have stayed the same concentration, as the drop of dye does not rise or fall. Therefore the onion cells and the solution are in equilibrium ✓.

ⓔ **1/2 marks awarded** 1 mark is awarded for understanding that the solution is in equilibrium with the onion cells but there is no reference to water potential for the second mark.

**(d) (i)** This stops any water evaporating.

**(ii)** They do not need to find the percentage change in mass.

ⓔ **0/3 marks awarded** Although the bung does stop evaporation, student B does not explain that this would change the concentration or water potential of the solution. In (d)(ii) student B is completely wrong, and has not seen that a change in density of the solution does not require the onion tissue to be exactly uniform in each tube.

## Question 7

Figure 6 shows the effect of substrate concentration on the rate of an enzyme-controlled reaction.

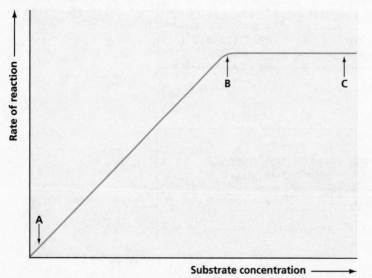

**Figure 6**

**(a)** Explain what is limiting the rate of reaction between:

   **(i)** A and B       (2 marks)

   **(ii)** B and C      (2 marks)

**(b)** Pravastatin is a drug used to treat high blood cholesterol concentration. It is similar in shape to the substrate of an enzyme in the liver that synthesises cholesterol. Explain how this drug is effective in reducing blood cholesterol concentration.    (3 marks)

Fresh pineapple contains a protein-digesting enzyme. Jelly contains gelatine, which is a protein. A student carried out an investigation to find the effect of pineapple on gelatine. She put fresh pineapple into a liquidiser, to produce a puree. She set up tubes as shown in Table 6. The tubes were placed in a refrigerator for 3 hours and then examined. The gelatine in tube 1 had not set, but the gelatine in tubes 2 and 3 had set.

| Tube number | Contents of tube |
|---|---|
| 1 | 6 cm³ gelatine + 2 cm³ pineapple puree + 2 cm³ water |
| 2 | 6 cm³ gelatine + 2 cm³ pineapple puree + 2 cm³ hydrochloric acid |
| 3 | 6 cm³ gelatine + 2 cm³ boiled pineapple puree + 2 cm³ water |

Table 6

**(c)** Explain why 2 cm³ of water was added to tubes 1 and 3, but not to tube 2.    (1 mark)

**(d)** **(i)** Explain the results for:

    ■ tube 1      (1 mark)

    ■ tube 2      (2 marks)

   **(ii)** Explain why tube 3 was necessary.    (2 marks)

ⓔ This question starts with a simple test of recall but then goes on to give you data from an investigation that you have probably not carried out. As always, you should read the information carefully and then apply what you have learned to this new situation.

**Student A**

**(a)** **(i)** Substrate concentration ✓ because when you add more substrate the rate of reaction increases ✓.

   **(ii)** Enzyme availability ✓ because all the active sites are being used ✓.

ⓔ **4/4 marks awarded** Student A has identified the limiting factor in each case, and given a clear reason.

**(b)** The drug is a competitive inhibitor ✓ that fits in the active site of the enzyme ✓. This means that fewer enzyme molecules will form an enzyme–substrate complex ✓, so less cholesterol is made.

ⓔ **3/3 marks awarded** Note that student A uses the technical terms 'active site' and 'enzyme–substrate complex'.

**(c)** To make the concentrations of enzyme and substrate the same in all the tubes ✓.

ⓔ **1/1 mark awarded** This is a well-expressed answer.

**(d) (i)** ■ Tube 1: The protease enzyme in the pineapple had digested the gelatine ✓.
■ Tube 2: The hydrochloric acid had denatured the enzyme ✓, so it could not digest the gelatine ✓.

**(ii)** To show that it is an enzyme in the pineapple puree that digests the gelatine ✓, because boiling the pineapple puree denatures the enzyme ✓.

ⓔ **5/5 marks awarded** Full marks for correct, clearly written answers.

## Student B

**(a) (i)** Substrate concentration ✓, because when you add more substrate the rate of reaction increases ✓.

**(ii)** Availability of active sites ✓, because all the active sites are being used ✓.

ⓔ **4/4 marks awarded** Student B has identified the limiting factor in each case, and given a clear reason.

**(b)** The drug is the same shape as the substrate, so it fits in the active site of the enzyme ✓. This means that fewer enzyme molecules will bind to an active site ✓, so less cholesterol is made.

ⓔ **2/3 mark awarded** The drug is a similar shape to the substrate, not the same shape, and student B has not described it as a competitive inhibitor.

**(c)** To make the volume in each tube the same.

ⓔ **0/1 mark awarded** The *concentrations* of the enzyme and substrate need to be the same.

**(d) (i)** ■ Tube 1: The pineapple had broken down the gelatine.
■ Tube 2: It was too acid, so it could not digest the gelatine ✓.

**(ii)** as a control

ⓔ **1/5 marks awarded** For tube 1 student B has not said that it is an enzyme in pineapple that digests the gelatine. For tube 2 student B recognises that the gelatine is still present, but does not relate this to the denaturation of the enzyme. In (d)(ii), although this is a control, student B has not shown that they understand the purpose of the control. The answer should say that it is to show that the enzyme is responsible for the effects noticed, and not any other factor.

# Question 8

Mitochondria and chloroplasts contain their own DNA, which is like bacterial DNA. They also contain ribosomes that are like bacterial ribosomes.

**(a)** Describe one way in which the following features of chloroplasts and mitochondria are different from those of eukaryotic cells:

    **(i)** DNA (1 mark)

    **(ii)** Ribosomes (1 mark)

A biologist wanted to obtain a sample of chloroplasts for an investigation. He homogenised some fresh spinach leaves in ice-cold, isotonic buffer solution. He strained the mixture through muslin, then spun the mixture in a test tube in an ultracentrifuge.

**(b)** **(i)** Explain why the spinach leaves were homogenised in a buffer solution. (2 marks)

    **(ii)** Explain why the buffer solution was:

      ■ ice cold (1 mark)

      ■ isotonic (1 mark)

**(c)** **(i)** The mixture was filtered before it was centrifuged. Suggest why. (1 mark)

    **(ii)** The biologist could obtain a fairly pure sample of chloroplasts from the mixture that had been spun in an ultracentrifuge. Explain how. (3 marks)

ⓔ This question is about a technique you should be familiar with. Therefore this is testing knowledge and understanding.

---

**Student A**

**(a)** **(i)** This will be circular instead of being linear ✓.

    **(ii)** These will be smaller than eukaryotic ribosomes ✓.

---

ⓔ **2/2 marks awarded** Full marks for well-expressed, correct answers.

**(b)** **(i)** This keeps the pH constant ✓ so that the enzymes do not denature ✓.

    **(ii)** ■ Ice cold: to stop enzyme activity ✓.

      ■ Isotonic: to stop the organelles taking in water by osmosis and bursting ✓.

ⓔ **4/4 marks awarded** These clear and accurate answers are worth full marks.

---

**(c)** **(i)** This gets rid of cell debris ✓.

    **(ii)** The pellet will contain nuclei as these are the densest organelles ✓. The supernatant will be put in a new tube and spun again ✓. The chloroplasts will be in the second pellet ✓. They can be mixed with isotonic buffer and used in an investigation.

---

ⓔ **4/4 marks awarded** In (c)(i) student A uses good terminology. Part (c)(ii) correctly refers to the density of the organelles, rather than the weight or size.

**Student B**

**(a) (i)** The DNA is circular.

**(ii)** They are smaller.

ⓔ **0/2 marks awarded** Student B has not made it clear whether they are referring to the feature of prokaryotic cells or eukaryotic cells.

**(b) (i)** To make sure the pH doesn't change ✓.

**(ii)** ■ Ice cold: to stop bacteria growing ✗.
■ Isotonic: to stop the cells bursting ✗.

ⓔ **1/4 marks awarded** 1 mark is awarded for the buffer keeping the pH constant but there is no reference to stopping the enzymes from denaturing. In (b)(ii) it does not matter whether the cells burst, as we want the cells to be broken open, which is why they are homogenised. It is the organelles that need to be kept intact.

**(c) (i)** To get rid of large particles ✗.

**(ii)** By centrifuging the mixture, so the heavy chloroplasts are in the pellet ✗.

ⓔ **0/4 marks awarded** In (c)(i) student B does not say how large particles might be in the tube. It needs to be clear that large particles will be parts of the tissue that have not been broken up in homogenisation. The term 'cell debris' is the best way to explain this. In (c)(ii) student B needs to explain that centrifugation separates organelles by density, and show that they understand that the chloroplasts will be in the second pellet.

# Question 9

**(a) (i)** Give two functions of proteins in cell-surface membranes. (2 marks)

**(ii)** Explain the features of phospholipids that enable them to form a bilayer in a cell-surface membrane. (2 marks)

In an investigation, scientists placed human cells in a solution of calcium ions. At regular intervals they measured the concentration of calcium ions in the solution and inside the cells. Their results are shown in Figure 7.

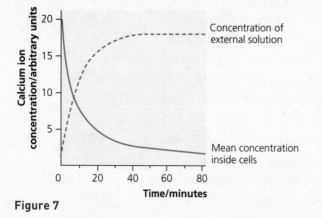

**Figure 7**

**(b)** By what process did the calcium ions leave the cells? Use evidence from Figure 7 to support your answer. (2 marks)

ⓔ This question starts with easy recall but then gives you a graph to test whether you can apply your knowledge correctly.

**Student A**

**(a) (i)** channel proteins ✓ and receptors ✓

**(ii)** The 'heads' are hydrophilic and the 'tails' are hydrophobic ✓. This means that the heads always arrange themselves towards water and the tails arrange themselves away from water ✓.

ⓔ **4/4 marks awarded** Clear and correct answers, for full marks.

**(b)** Active transport ✓, because the ions reach a much higher concentration outside the cell than inside, so they move against a concentration gradient ✓.

ⓔ **2/2 marks awarded** Full marks again for a clear, well-expressed answer.

**Student B**

**(a) (i)** carrier proteins ✓ and antigens ✓

**(ii)** The 'heads' are water loving and the 'tails' are water hating ✗.

ⓔ **2/4 marks awarded** In (a)(ii) student B should have used the terms 'hydrophilic' and 'hydrophobic', and then explained how these properties cause the phospholipids to orientate themselves into a bilayer.

**(b)** Active transport ✓, because the ions leave the cell against a concentration gradient ✓.

ⓔ **2/2 marks awarded** A succinct answer is all that is needed for full marks.

## Question 10

**(a)** Describe how B cells respond to a pathogen. (6 marks)

**(b)** Explain how a vaccine can prevent a person from developing an infectious disease. (5 marks)

**(c)** It is not necessary for everyone in a population to be vaccinated against a disease for the disease to be rare or even eradicated. Explain why. (4 marks)

ⓔ This type of question tests your ability to write in continuous prose.

**Student A**

**(a)** The antigen is presented to a B cell in the lymph nodes ✓ that has receptors in its membrane that are complementary to the antigen ✓. This B cell is activated by a helper T cell ✓, so it divides to form two clones. One clone consists of plasma cells ✓, which secrete antibodies specific to the antigen ✓. Another clone is memory B cells, which retain an immunological memory of the original antigen ✓.

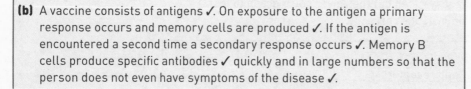

**6/6 marks awarded** This is an excellent account, which is detailed and clearly expressed.

> **(b)** A vaccine consists of antigens ✓. On exposure to the antigen a primary response occurs and memory cells are produced ✓. If the antigen is encountered a second time a secondary response occurs ✓. Memory B cells produce specific antibodies ✓ quickly and in large numbers so that the person does not even have symptoms of the disease ✓.

**5/5 marks awarded** This is another well-structured response.

> **(c)** There is a principle called herd immunity ✓. If most people in a population are vaccinated they cannot catch the disease ✓. People who are not vaccinated can still get the disease, and could also die if it is a serious disease ✓. But there are very few unvaccinated people in the population they can pass the infection on to ✓.

**4/4 marks awarded** This shows good use of technical vocabulary.

**Student B**

> **(a)** The antigen is presented to a B cell in the lymph nodes ✓, which is activated by a helper T cell ✓. It divides to form a clone consisting of plasma cells ✓, which secrete antibodies specific to the antigen ✓. It also produces memory B cells, which remember the original antigen ✓.

**5/6 marks awarded** A further mark would have been awarded for reference to the receptors in the B cell membrane that are complementary to the antigen.

> **(b)** The vaccine causes plasma cells and memory cells to be produced ✓. If the antigen is encountered again the memory B cells produce specific antibodies ✓ so fast that the person does not become ill ✓.

**3/5 marks awarded** Student B does not explain that the vaccine contains antigens, and there is no reference to the primary and secondary response.

> **(c)** People who are vaccinated cannot become infected with the pathogen that causes the disease ✓. This means that infected people are unlikely to meet an unvaccinated person to pass on the infection ✓.

**2/4 marks awarded** The answer needs to mention herd immunity, and there is no explanation of the concept, i.e. that if most people in a population are vaccinated, it is difficult for an infectious disease to spread.

# Required practical answers

## Required practical 1

**1** The investigation should be repeated using boiled potato discs (to ensure the enzyme is denatured) but keeping all other factors the same.

**2** $rate = \dfrac{volume\ of\ oxygen\ evolved}{time}$

   **(a)** $4.3/20 = 0.215\,cm^3\,s^{-1}$

   **(b)** $0.2/20 = 0.01\,cm^3\,s^{-1}$

**3** Repeat the investigation, putting the tube in a water bath at different temperatures each time, for example 0, 5, 10, 15, 20, 25, 30, 35, 40, 45 and 50°C. Keep everything else the same, for example, volume and concentration of hydrogen peroxide used and number of potato discs. Find the volume of oxygen evolved at each temperature in the same time, for example 120 seconds. Plot a graph of temperature ($x$-axis) against rate of reaction ($y$-axis). The optimum temperature is where the rate of reaction is fastest. It would be even better to repeat this again at smaller temperature intervals around the temperature value where the rate of reaction is highest, to get a more accurate value for the optimum temperature.

**4** Repeat the investigation using the same amount of potato and the same volume of hydrogen peroxide solution, but varying the concentration of hydrogen peroxide. Find the volume of oxygen evolved in a fixed time period, for example 120 seconds, and calculate the rate of reaction for each substrate concentration. Plot a graph of hydrogen peroxide (substrate) concentration on the $x$-axis and the rate of reaction on the $y$-axis.

**5** This would give a more reliable mean value as it reduces the effect of outliers.

## Required practical 2

**1** You would keep a tally of how many cells are in each stage. You can find the percentage of time spent in each stage by the formula:

$$\dfrac{number\ of\ cells\ in\ stage}{total\ number\ of\ cells} \times 100\%$$

Then you find how long it takes the cell to complete mitosis and work out how long each stage is by using the percentages calculated above.

## Required practical 3

**1** $\dfrac{final\ mass - original\ mass}{original\ mass} \times 100\%$

**2** So that the tubes could be compared. The mass of potato in each dish is not exactly the same.

**3** 1.0 M sucrose has a lower water potential than the potato cells, so water leaves the potato cells by osmosis. Therefore they lose mass. The 0.1 M sucrose has a higher water potential than the potato cells, so water enters the potato cells by osmosis, causing the potato cells to gain mass.

**4** There are several possible answers here but one possibility is to compare the intensity of colour to a colour chart.

# Knowledge check answers

## Knowledge check answers

**1** The same atoms are arranged differently to give two different sugars.

**2** ionic and hydrogen bonds, and disulfide bridges

**3 a** ester, **b** peptide

**4** It is not the same shape because if the active site was the same shape as the substrate they would not fit together. The student should have said that they have *complementary* shapes.

**5** Sucrase has a complementary active site for sucrose to fit in. Maltose is a different shape and will not fit in the active site.

**6** hydrogen bonds and ionic bonds

**7** A buffer solution keeps the pH constant. This is necessary because temperature is the independent variable, so pH must be kept constant.

**8** The line would level off and fall if substrate concentration was limited.

**9** The line must be below the original line with an initial gradient that is less steep. It then becomes horizontal, but at a lower level than the 37°C line. It should be labelled as 25°C.

**10** The line should be below the competitive inhibitor line and to the right. It should still meet the 'without inhibitor' line, although further to the right.

**11** The drug will fit in the active site of aldehyde oxidase. This means that fewer enzyme active sites will be available to break down acetaldehyde. Therefore acetaldehyde will accumulate, making the person feel nauseous, so they will be less likely to drink excessively.

**12** The line should be the same shape as the non-competitive inhibitor curve, but it will have a lower gradient and will level off below the non-competitive inhibitor curve.

**13** If 20% contain thymine, 20% will contain adenine as these bases are complementary. There are 60% of nucleotides left, so 30% will contain cytosine and 30% will contain guanine.

**14** Similarities: both contain phosphate; both may contain adenine, cytosine or guanine. Differences: DNA contains deoxyribose but RNA contains ribose; DNA nucleotides may contain thymine to pair with adenine, but RNA has uracil instead.

**15** Only the complementary bases can bind to the bases on the original/old (template) strand, so the new strand will always be an exact but complementary copy of the old strand.

**16 a** Similarities: it contains ribose; it contains adenine, which is one of the bases found in RNA nucleotides. Difference: it has three phosphate groups attached, instead of one.

**17 b** They have a negative charge that attracts the positive charges on water molecules, which form a 'shell' around them.

**18 a** Enzymes are proteins, so pancreas cells have a lot of RER where the enzymes will be synthesised and then transported.

**b** These will contain more mitochondria because wing muscles are very active and therefore need a lot of ATP. Mitochondria produce ATP in aerobic respiration.

**19** Because it is made of many similar cells that carry out a specific function.

**20** DNA helicase and DNA polymerase

**21** Plant cell: scale bar measures 8 mm and represents 5 μm

$$\text{magnification} = \frac{\text{measured size}}{\text{actual size}}$$

$$= \frac{8000}{5}$$

$$= 1600$$

Bacterial cell: scale bar measures 12 mm and represents 0.1 μm

$$\text{magnification} = \frac{\text{measured size}}{\text{actual size}}$$

$$= \frac{12000}{0.1}$$

$$= 120000$$

**22** Use a fresh sample of leaves and homogenise them in ice-cold isotonic buffer. Strain through muslin to remove cell debris, then centrifuge in an ultracentrifuge. The first pellet is likely to contain nuclei, as these are the densest organelles. Pour the supernatant into a fresh tube and put in an ultracentrifuge and spin again. The second pellet will contain chloroplasts. Pour off the supernatant and resuspend the chloroplasts in isotonic buffer.

**23 a** The distance between sister chromatids.

**b i** 10 minutes **ii** Because that is when the chromatids start to separate and move towards the poles of the cell.

**24** It will stop the cells at prophase. The cells will not be able to enter metaphase, as this needs a spindle to be present.

**25** Cancer cells divide quickly and repeatedly by mitosis, so inhibiting mitosis helps to treat cancer. If the DNA cannot unwind then it cannot replicate, so mitosis cannot happen.

**26** This is useful to kill bacteria because bacterial cells cannot divide unless the DNA has replicated. However if the enzyme is not present in human cells, this means that the drug cannot harm human cells.

**27** They do not have a cell structure. They show no signs of life such as respiration. They have no metabolism of their own and can only reproduce inside a host cell.

**28** It is an ion and carries a charge, making it soluble.

**29** These proteins have a specific tertiary structure so that only one kind of molecule is able to pass through them.

**30** A enters by simple diffusion, as the rate of diffusion into the cell increases as the concentration of the molecule outside the cell increases. B enters by facilitated diffusion because the rate of diffusion into the cell increases as the concentration of the molecule outside the cell increases, until a point is reached where the rate levels off. This is the point where all the carrier proteins are being used, so diffusion cannot go any faster.

**31 a**

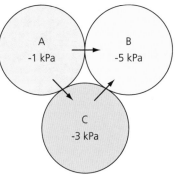

**b** The fastest rate of flow is from A to B as the water potential gradient is greatest between these two cells.

**32 a** It will take in water and its volume will increase. There is no cell wall around it, so once the cell reaches a certain volume it will burst.

**b** A plant cell will not burst as it has a fairly rigid cell wall around it, so there is a limit to how much its volume can increase when it takes in water by osmosis.

**33** Ion A enters by diffusion because it enters the cell down a concentration gradient. Ion B enters by active transport, as there is a greater concentration of the ion inside the cell than outside, so it is entering the cell against a concentration gradient.

**34** This increases the surface area of the cell, so that digested food molecules can be absorbed more efficiently.

**35** It actively transports sodium ions out of the cell, creating a low concentration of sodium ions inside the cell. Therefore glucose diffuses in at B, carrying a glucose molecule with it.

**36** The kidney will have antigens on it that are 'self' to the person who donates the kidney but are 'non-self' to the recipient. Therefore the recipient's immune system will mount an immune response against the kidney unless the immune system is suppressed.

**37** This is because they are all the same and are specific to the same antigen.

**38** This is a 'booster' to increase the number of memory cells present.

**39** After a vaccine is administered, it takes time for a primary response to occur and for specific antibodies and memory cells to be produced. If the person is already infected, a primary response will be happening in their body anyway. Vaccines are only effective if a primary response takes place before the person is infected, so that a rapid secondary response is stimulated when the person encounters the pathogen.

**40** This is passive immunity because the baby is receiving ready-made antibodies and does not produce memory cells of its own.

**41** This is active immunity because the person is mounting their own immune response to an antigen, making their own antibodies and memory cells.

**42** The antibodies would treat the tetanus infection, but the horse antibodies would act as a non-self antigen to the person. Therefore the person will mount an immune response against the horse antibodies. If they are used a second time, there will be a big secondary response, with large numbers of antibodies released to bind to the horse antibodies, and they would not be effective.

**43** Helper T cells stimulate antibody production by plasma cells, so fewer antibodies will be made in response to pathogens.

**44** HIV has an attachment protein that fits into a receptor protein on the helper T cell membrane. No other kind of cell has this receptor protein on its cell surface membrane.

**45** Mouse antibodies would be slightly different in their structure from human antibodies. A human given normal mouse antibodies would mount an immune response against them and the antibodies would not be effective.

**46 a** This is so that any antibodies that have not bound to the antibodies in the well are washed away. If they remained in the well they could give a false positive result.

**b** The enzyme causes a colour change when the final solution is added. This is necessary so that a visible colour change is seen and the doctor can tell whether the test is positive or negative.

# Index

# Index